Julio Surreaux Chagas

Balanced Vibratory Conveyor Design

Julio Surreaux Chagas

Balanced Vibratory Conveyor Design

Application: Food, chemical, pharmaceutical, smoke and other industries

ScienciaScripts

Imprint
Any brand names and product names mentioned in this book are subject to trademark, brand or patent protection and are trademarks or registered trademarks of their respective holders. The use of brand names, product names, common names, trade names, product descriptions etc. even without a particular marking in this work is in no way to be construed to mean that such names may be regarded as unrestricted in respect of trademark and brand protection legislation and could thus be used by anyone.

Cover image: www.ingimage.com

This book is a translation from the original published under ISBN 978-613-9-77737-2.

Publisher:
Sciencia Scripts
is a trademark of
Dodo Books Indian Ocean Ltd. and OmniScriptum S.R.L publishing group

120 High Road, East Finchley, London, N2 9ED, United Kingdom
Str. Armeneasca 28/1, office 1, Chisinau MD-2012, Republic of Moldova, Europe
Printed at: see last page
ISBN: 978-620-6-60141-8

PREFACE

I am a mechanical engineer who graduated in 1966 from the School of Engineering of the Federal University of Rio Grande do Sul - UFRGS, Brazil. I worked in industries, basic sanitation, petrochemicals and oil refineries for 44 years, from 1967 to 2011, in the areas of maintenance, planning, projects and engineering works.

At Souza Cruz S.A. I worked at the Porto Alegre Cigarette Factory and in the Engineering Department of the Head Office in Rio de Janeiro from 1967 until 1984. From 1975 I worked in the Engineering Department in Rio de Janeiro as a Senior Project Engineer advising the 8 cigarette factories on increasing production and modernising tobacco processing.

One of my activities was to design and monitor the manufacture of balanced vibrating conveyors, also known as "Panda" conveyors. Around 60 vibrating conveyors were manufactured to meet the needs of cigarette factories.

I'm happy for the memories of the work I've done in my life. I'm grateful to the superiors, colleagues and subordinates I've worked with and learnt from in my profession and in life.

Porto Alegre, 1st March 2019

Mechanical Engineer Julio Surreaux Chagas

SUMMARY

CHAPTER 1

INTRODUCTION

The first industry to widely use balanced vibratory conveyors in the manufacturing process that we know of was "The Imperial Tobacco Co. of Canada Ltd.", Montreal, Que.

From 1956 to 1965, they installed more than 100 balanced vibratory conveyors in their industrial plants to replace unbalanced vibratory conveyors and belt conveyors with the aim of reducing maintenance costs and making the equipment easier to clean and sanitise. Balanced vibratory conveyors are usually more expensive than belt conveyors.

At that time, the cigarette factories of Cia. Souza Cruz S.A. in Brazil used very rudimentary vibratory conveyors without balancing, with a stainless steel conveyor chute supported by springs made from the wood of the Genipapo tree. They were noisy due to the vibration of the whole assembly and frequently broke down and broke the springs. The vibrating conveyors were normally used for feeding and discharging drying cylinders and packaging tobacco.

In 1975, the Board of Directors of Souza Cruz S.A. approved the increase and modernisation of production at the eight cigarette factories. One of the decisions was to replace most of the existing tobacco processing conveyors in the cigarette factories with balanced vibratory conveyors according to the model developed by The Imperial Tobacco of Canada Ltd. This company belongs to the British-American Tobacco Company Limited group with headquarters in London, together with Cia Souza Cruz S.A. with headquarters in Rio de Janeiro. Over a period of five years, around 60 balanced vibratory conveyors were installed in the cigarette factories of Cia. The design of these conveyors was standardised to facilitate manufacture and reduce costs.

CHAPTER 2

GENERALISATIONS

2.1 - OBJECTIVE

The aim of this manual is to provide engineers with knowledge on the design of balanced vibratory conveyors for use in the food, chemical, pharmaceutical, tobacco and other industries.

2.2 - DEFINITION

It is a device for transporting bulk materials using a vibrating chute with elastic springs fixed to a base. The vibration is produced by an eccentric driven by an electric motor. Below the chute is a steel profile frame with elastic springs fixed to the base with reverse vibrations of the chute. The weight of the rail must equal the weight of the counterweight to balance the system.
A vibrating conveyor in operation is balanced when the rail springs and counterweight are vibrating at their natural frequency (VPM), coinciding with or as close as possible to the rotation of the eccentric. This results in minimal vibration at the base, quiet operation and minimal maintenance. By varying the rotation of the eccentrics on a vibrating conveyor, you can tell when the springs are naturally vibrating.
The use of imported fibreglass springs (Epoxy Glass) manufactured by the 3M Company with the following dimensions was standardised:

       Lengths: 12"   -12   "
       Widths:2   " -2"
       Thicknesses: 3/16" -1/4"

The free length for the springs to vibrate when mounted on the conveyor is 8 1/4" because 1 7/8" of each end is used to attach them to the cleats, as shown in fig. 25. Graph no. 1 below identifies the natural frequencies of the spring for each weight it is subjected to.

CHART NO. 1

<u>NATURAL FREQUENCY</u> FOR *FIBREGLASS* SPRINGS SPRING <u>DIMENSIONS</u>: 2" WIDE X 8" FREE LENGTH

Note: No relationship should be made with the drive springs

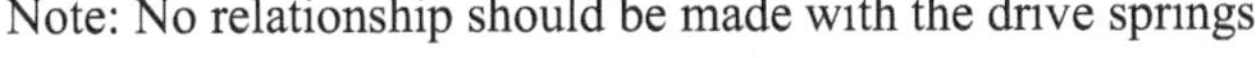

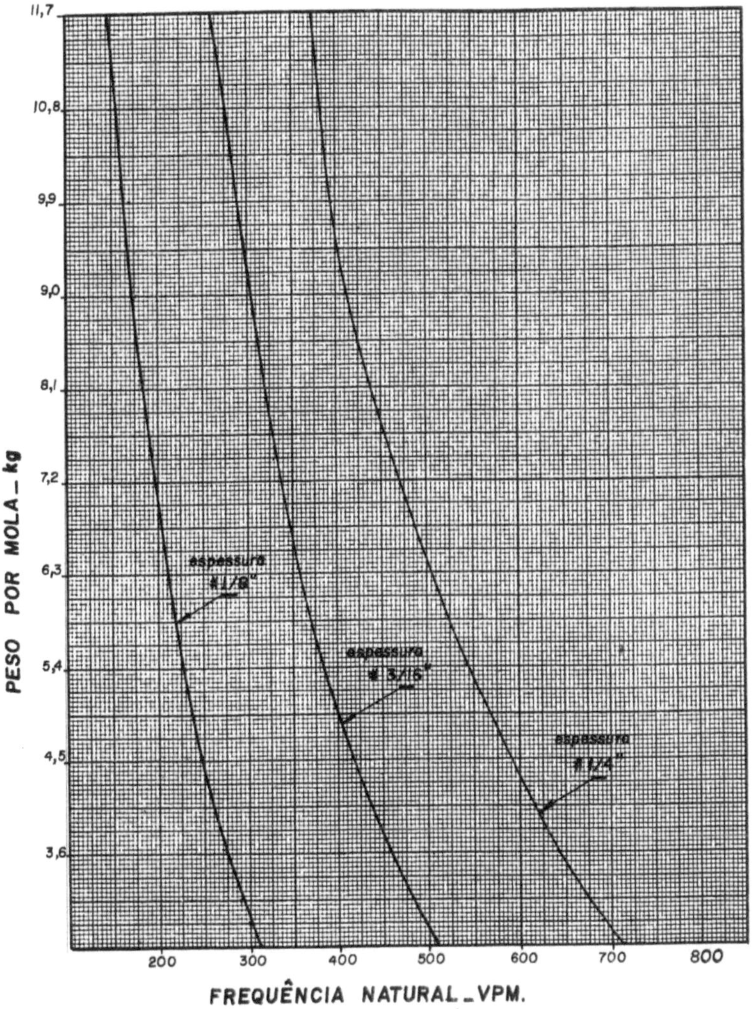

2.3 - FIELDS OF APPLICATION

They apply to the transport of the following products in bulk:
- Powders, grains, granules, pellets, sawdust, wood chips, dried CRS or stalk, cut tobacco, threshed tobacco, CRS with 30 per cent moisture or WTS with 48 per cent moisture;
- Density of the material (loose without tamping) below 300 kg/m3;
- Absolute humidity less than 48%.

CHAPTER 3

DESCRIPTION OF THE SUBSETS

Figure 1 shows a vibrating conveyor with its sub-assemblies:

FIGURE 1

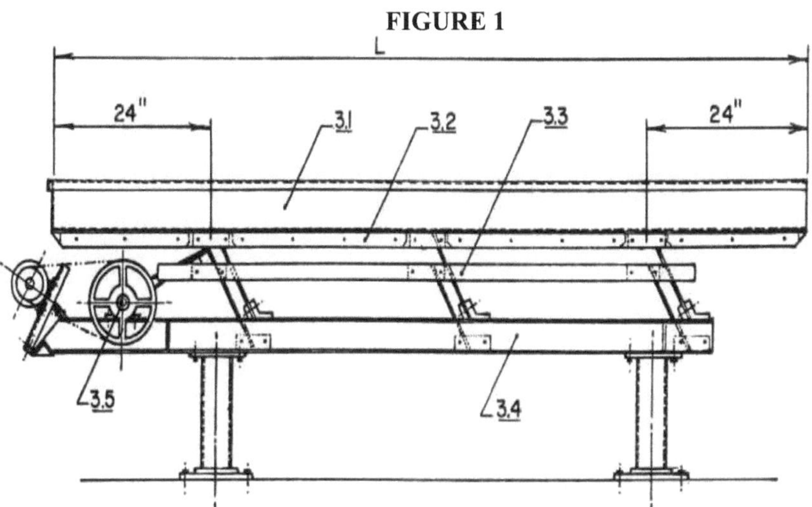

They are made up of the following subsets:
3.1 - Guttering
3.2 - Chassis
3.3 - Counterweight
3.4 - Base
3.5 - Drive

3.6 - CALHA
The use of AISI-304 n°20 stainless steel was standardised for the manufacture of the gutters.

The chute of a vibrating conveyor should be made as light as possible, and care should be taken in the design to avoid fitting unnecessary accessories, preferably on the fixed part of the conveyor (base). Examples of these accessories include inlet and outlet chutes and funnels.

3.1.1 - Types of gutter

The gutter section can have the following shapes: circular, rectangular closed or open; half-cane, part circular and part half-cane. However, the most common is the open rectangular section as shown in figure 2.

FIGURE 2

The standard widths adopted for manufacturing vibrating conveyor chutes are:
L (mm) = 480; 620; 760; 900; 1040.

The height (H) can be variable, but usually 150 mm or 200 mm is used. A conveyor with an open rectangular chute is considered to reach its maximum conveying capacity when the height of the product belt is 50 mm from the top of the chute, as shown in figure 3.

FIGURE 3

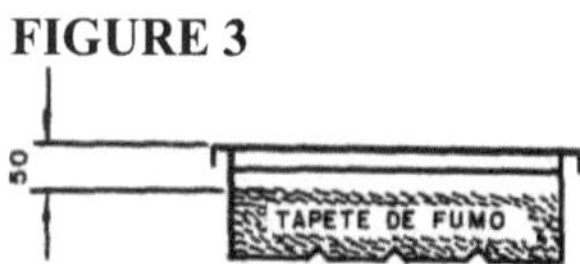

On conveyors with a circular chute, the maximum conveying capacity is reached when the product belt reaches half of the section, as shown in figure 4.

FIGURE 4

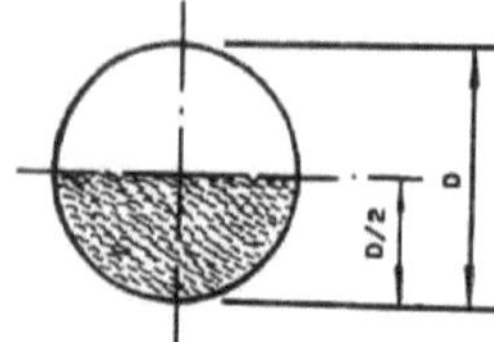

3.1.2 - Conveyor length

Obviously, the length of balanced vibratory conveyors is limited. Horizontal vibratory conveyors 13 metres long have been built without any operating problems. Imperial Tobacco Co. of Canada Ltd. has installed a conveyor 24 metres long with no operating problems. We believe that inclined conveyors are more limited in length than horizontal conveyors.

3.1.3 - Rectangular rail stiffness

The gutters must be manufactured in longitudinal sections and electrically welded. The open rectangular gutters are welded with U-shaped crosspieces of stainless steel sheet no. 20, spaced 1.50 metres apart, in order to increase their rigidity, as shown in figure 5.

FIGURE 5

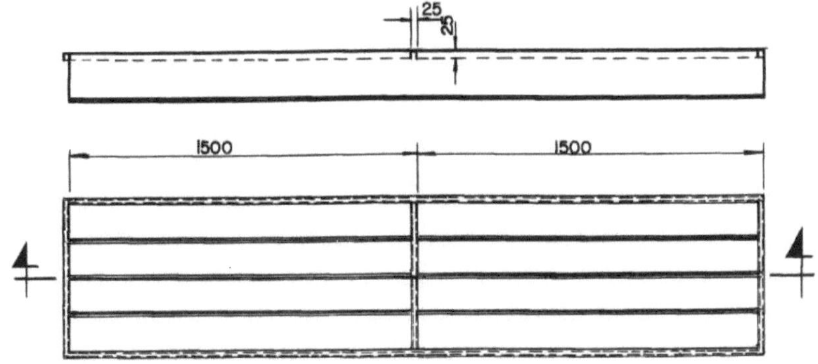

To increase the rigidity of the chute, making the vibratory conveyor quieter, the bottom plate of the chute should have bevelled edges, as shown in figure 6.

FIGURE 6

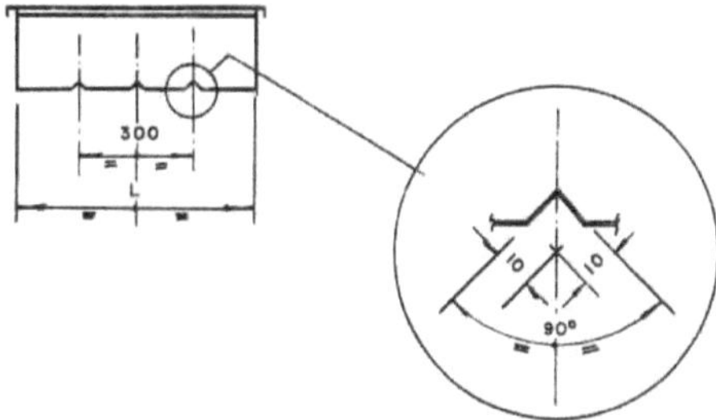

Due to the size limitations of No. 20 stainless steel sheets, open troughs with widths equal to or greater than 620 mm are made in 3 parts, with separate bottoms and sides, as shown in figure 7.

FIGURE 7

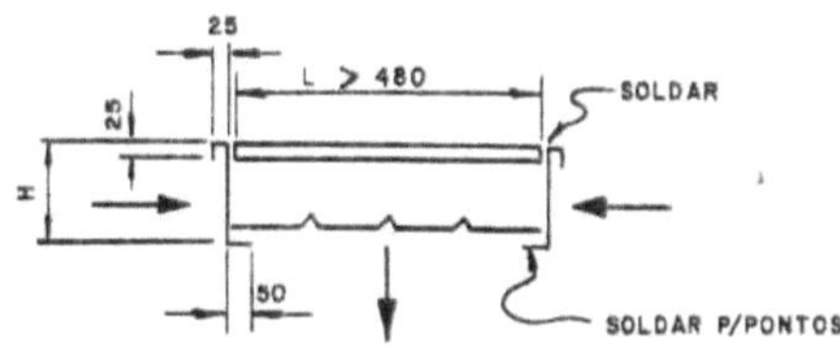

In the unique case of the rail with a width of 480 mm and a height of 150 mm, the rail is made in one piece.

3.1.4 - Fixing the rectangular rail to the chassis

Along the longitudinal length of the open rectangular gutter, in the space between the aluminium cleats, a folded stainless steel sheet (L) no. 20 measuring 2" x 2" is dotted at both ends of the bottom of the gutter, as shown in figure 8.

FIGURE 8

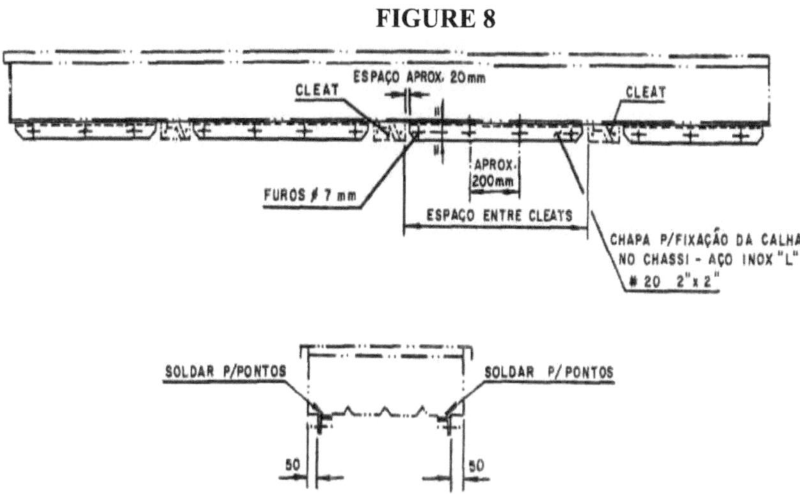

This longitudinal plate is fixed to the 2" x 2" x 1/4" aluminium angle bracket that belongs to the chassis sub-assembly. In the transverse direction, at the top of the aluminium cleats, L-shaped stainless steel plates n°20 measuring 15 mm x 50 mm are externally dotted at the bottom of the rail. These plates are fixed to the cleats, which belong to the chassis sub-assembly, using 1/4" x 5/8" diameter screws, as shown in figure 9.

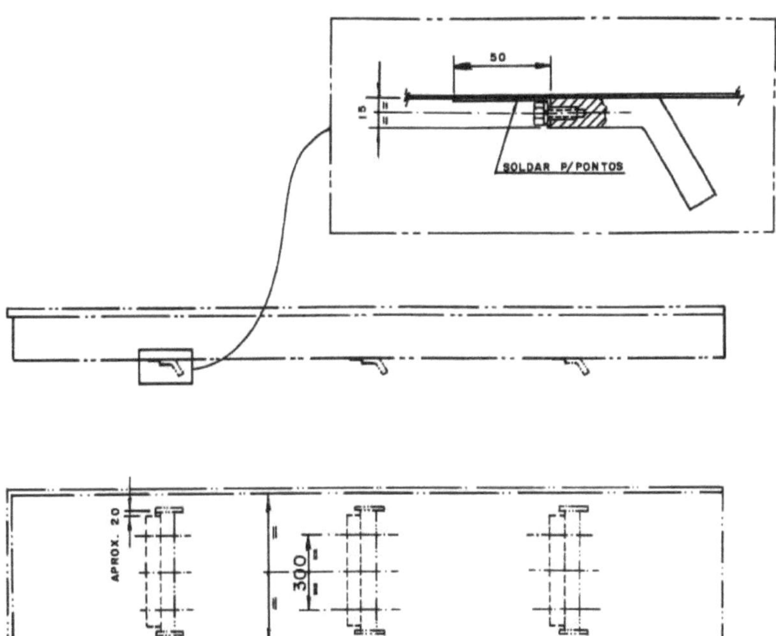

3.1.5 - Fixing the circular rail to the chassis

When the chute is circular, it must be attached to the chassis by means of pieces of stainless steel n° 18, with a length at the widest part of 300mm, spaced 1m apart along the longitudinal length. The aforementioned piece must be welded to the body of the cylinder and bolted to the 2" x 2" x 1/4" aluminium angle bracket, which belongs to the chassis sub-assembly, as shown in figures 10, 11 and 12.

FIGURE 10

DIÂMETRO DA CALHA	A
480	90
620	120
760	150
900	180
1040	210

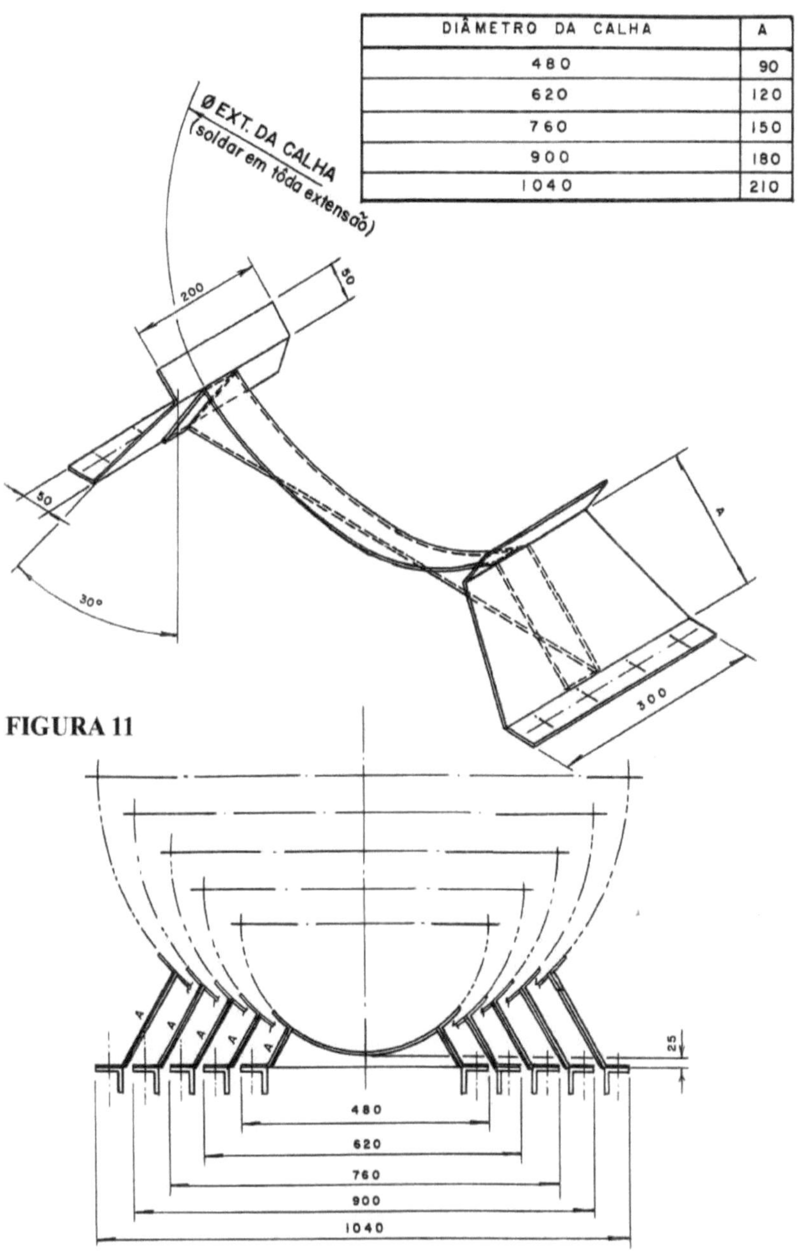

FIGURA 11

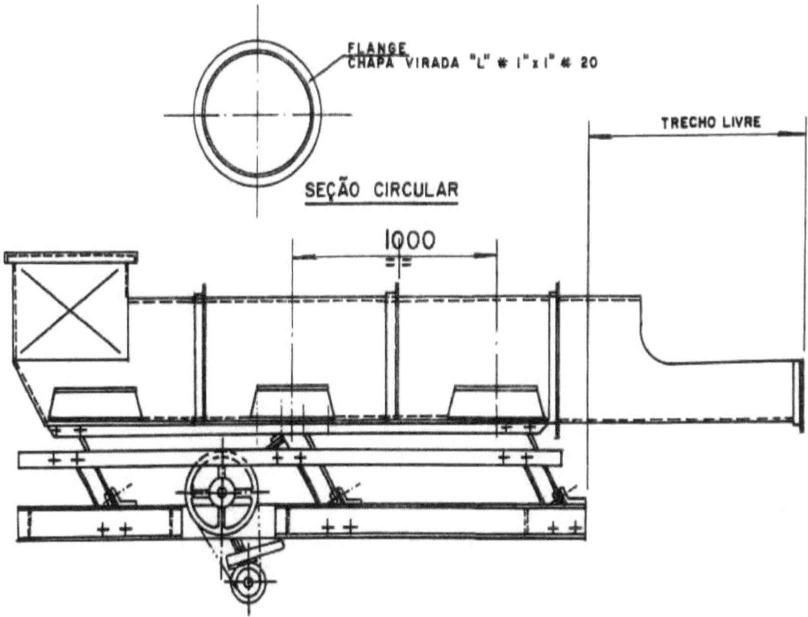

3.1.6 - Overhanging ends

It is recommended that the length (L) of the overhanging ends of vibratory conveyor chutes does not exceed 24", as shown in figure 1.

When the length of the overhanging ends exceeds 24", the vibrating conveyor makes more noise and the end springs are not balanced.

In inclined vibratory conveyors, it is common to make an initial section of the chute with an inclination of 5° to the horizontal and a length of no more than 24", as shown in figure 13.

On inclined conveyors, drains should be installed at the lower end of the chute to drain off water resulting from very wet product or from cleaning the chute, as shown in figures 14 and 15.

FIGURE 13

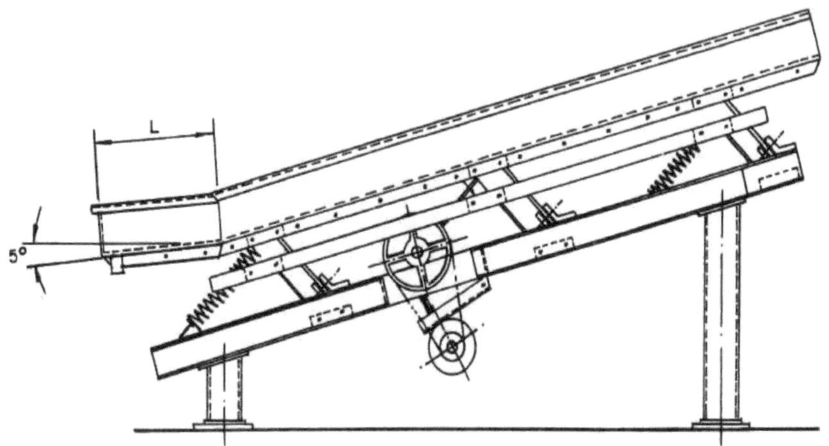

FIGURE 14

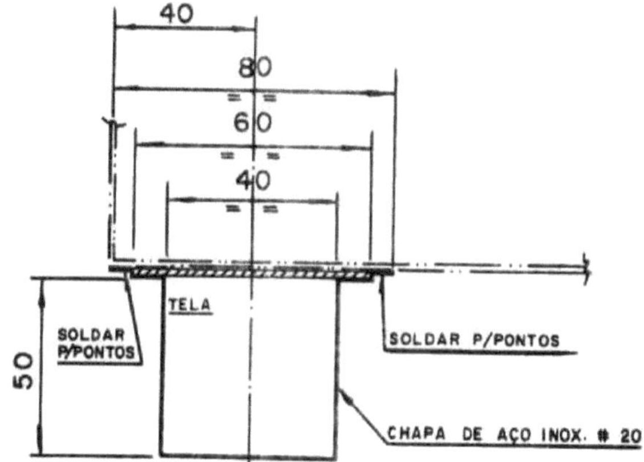

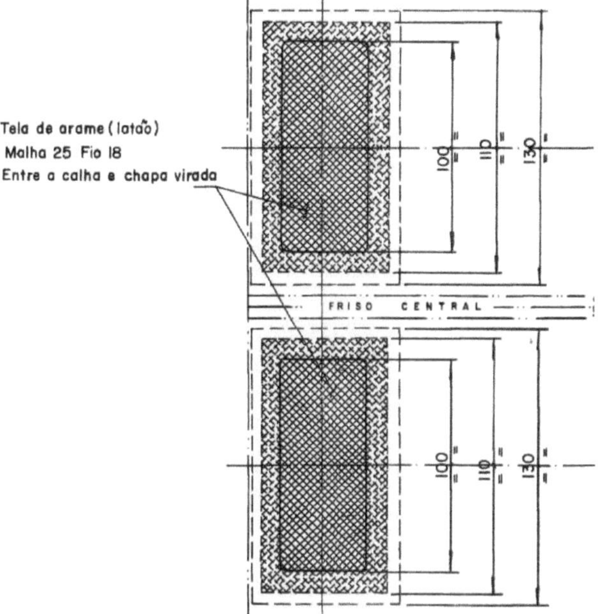

2 horizontal vibratory conveyors with the same direction of flow are connected as shown in figure 16.

FIGURE 16

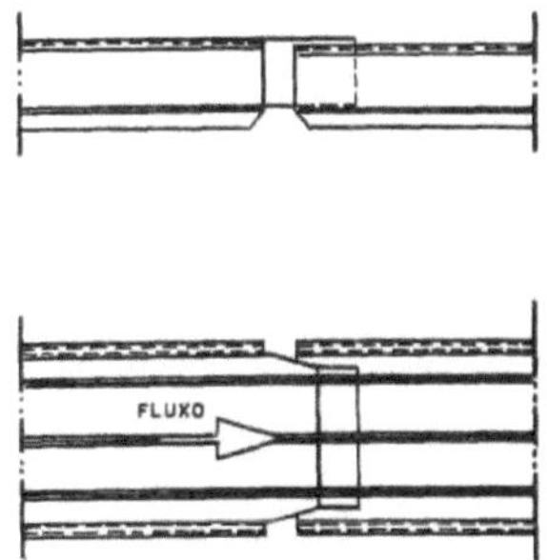

3.1.7 - Registers and discharge funnels

Some conveyors require the installation of registers with discharge funnels and internal divisions. Figures 17, 18, 19, 20, 21, 22 and 23 below show the various types of registers and drawers, and discharge funnels.

FIGURE 17

REGISTRO LADO ESQUERDO DA CALHA
PORCA TIPO BORBOLETA ∅1/4"NF C/ARR. DE PRESSÃO
PERFIL DE ALUMÍNIO "L" 2" x 2" x 1/4"
FUROS ∅ 7/32"
FURO ∅ 9/32"
20
40
4,0
40
25
R=2
70
15
REGISTRO ABERTO
150
25
25
100
SOLDAR
15
REGISTRO FECHADO
R=100
2
15°
AÇO INOX ∅14
100
15
IGUAL ALTURA DA CALHA
FIXAR C/PARAF. ∅3/16"W
65°
90°
10°
50
8
8
50
23
R=15
10
5
20
100
15
10
R=15
10
23
10
150
40
1"
∅9/32"
ALUMÍNIO
2 x 45°
20
20
50
40
30
CHAPA ∅1/8" LATÃO
∅9/32"
SOLDAR
PARAF. ∅1/4"W x 1 1/2"

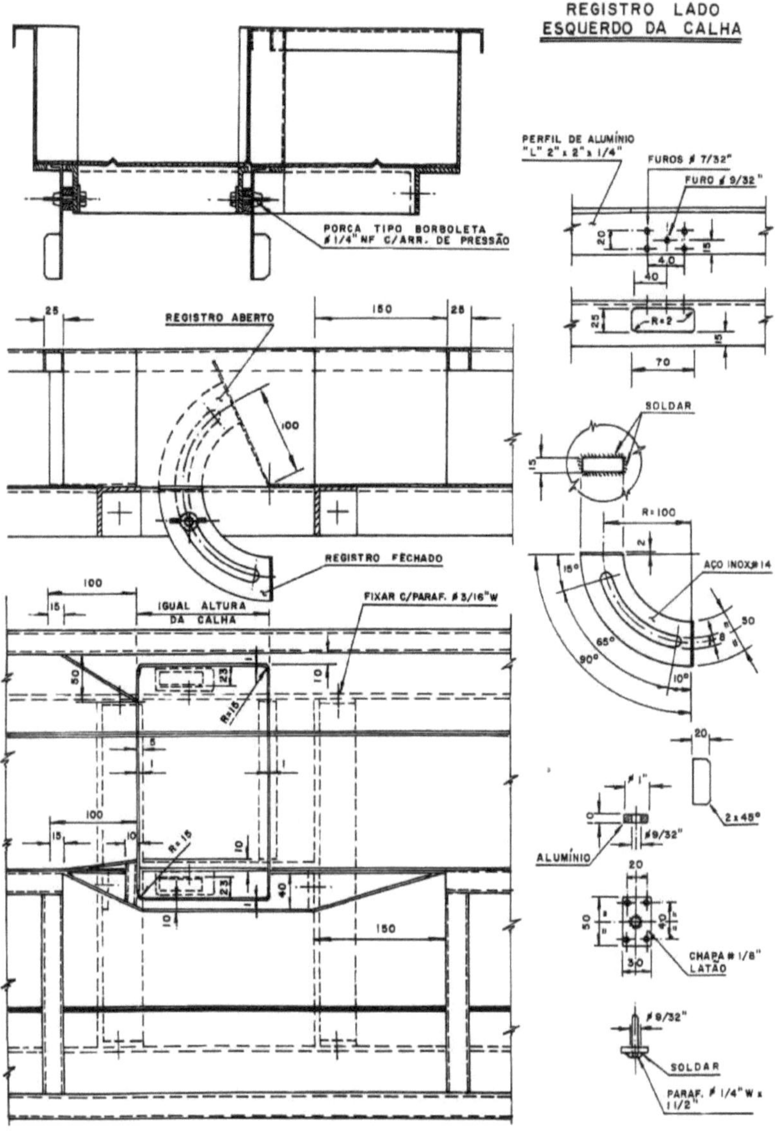

FIGURE 18

REGISTRO ABERTO
100
PARAF. CAB. BOLEADA
ø 3/16" W x 1/2" COMP.
C/ARRUELA DE PRESSÃO
E PORCA SEXTAVADA
REGISTRO FECHADO
LARGURA IGUAL ALTURA DA CALHA
100
15
50
23
10
R=15
DOBRADIÇA TIPO PIANO
AÇO INOX # 20
CHAPA DE AÇO INOX. # 14 SOLDADA
10
R=15
23
10
AÇO INOX. # 20
SOLDAR
FURO ø 9/32"
FURO ø 7/32"
40
20
30
50
PARAF. CAB. SEXT.
ø 1/4" W x 1 1/2" COMP.
LATÃO # 1/8"
ø 9/32"
10
ALUMÍNIO
SOLDAR
15
AÇO INOX. # 14
R=100
2
15°
20
2 x 45°
65°
90°
10°
5,0
ALUMÍNIO "L" 2"x 2"x 1/4"
FURO ø 9/32"
FURO ø 7/32"
42
40
12
15
25
15
R=2
70
REGISTRO CENTRO DA CALHA

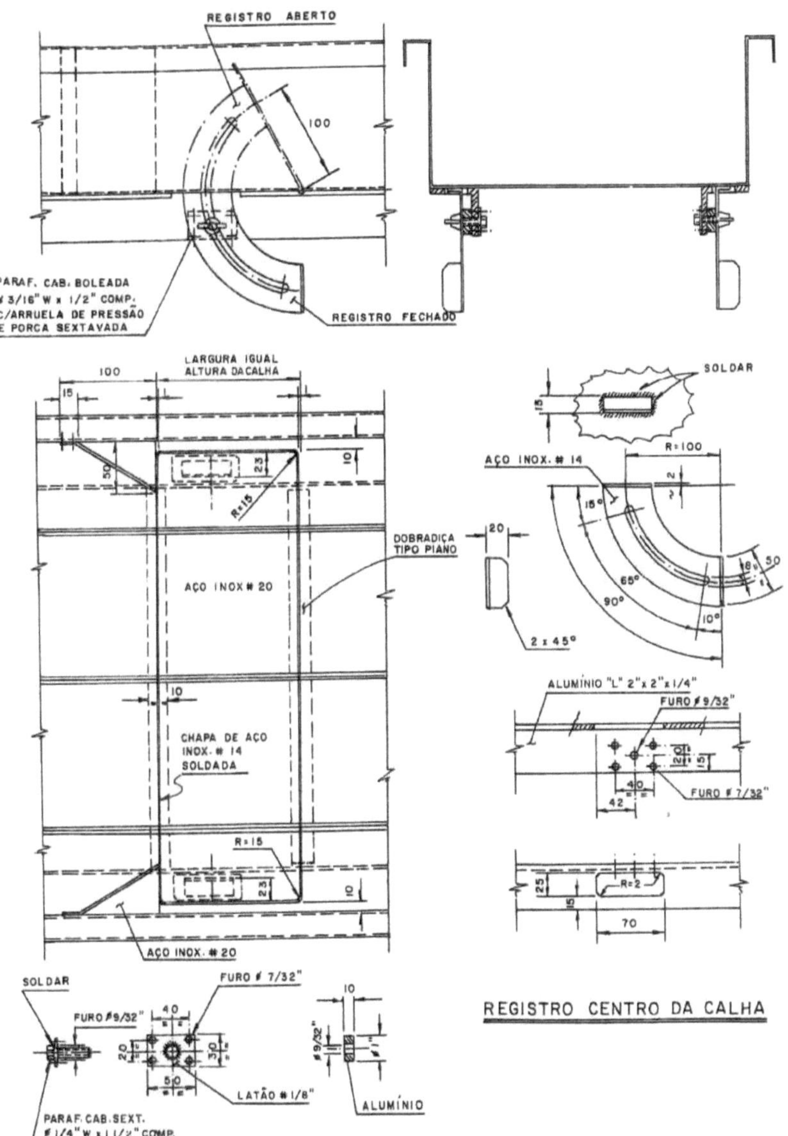

17

FIGURE 19

SOLDAR
R = 100
PERFIL DE ALUMÍNIO
"L" 2" x 2" x 1/4"
AÇO INOX # 14
2 x 45°
15°
75°
90°
10°
20
50
IGUAL ALTURA DA CALHA
REGISTRO ABERTO
AÇO INOX. # 20
100
FURO ⌀ 9/32"
FURO ⌀ 7/32"
40
40
70
R = 2
PERFIL DE ALUMÍNIO
"L" # 2" x 2" x 1/4"
PORCA TIPO BORBOLETA ⌀ 1/4" - W
COM ARRUELA DE PRESSÃO
REGISTRO FECHADO
FIXAR C/PARAF. ⌀ 3/16" W
ALUMÍNIO
⌀ 9/32"
CHAPA 1/8"
LATÃO
FURO ⌀ 7/32"
SOLDAR
PARAF. CAB. SEXT.
⌀ 1/4" W x
1 1/2" COMPR.
AÇO INOX # 20
R = 15
85
100
IGUAL ALTURA DA CALHA
150
15
5
40
100
R = 15
REGISTRO LADO
DIREITO DA CALHA

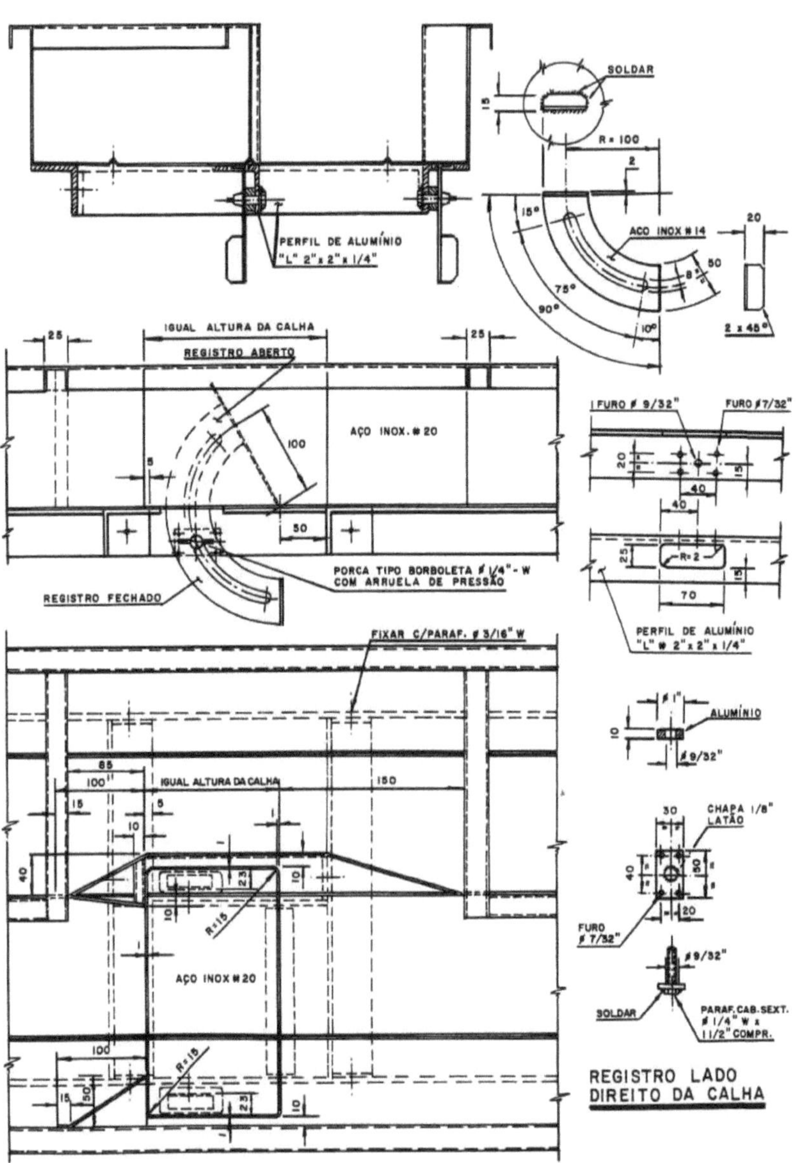

FIGURE 20

140
100
20
20
25
7/32"
40
CHAPA CORREDIÇA # 20
VEJA ABAIXO DET. DA FIXAÇÃO
40
100
100
15
150
25
50
25
240
25
20
140
200
PARAF. CAB. SEXT. ⌀ 3/16" W x
3/4" COMPR.
ARRUELA DE PRESSÃO
PORCA TIPO BORBOLETA
PONTEAR
SOMENTE UM
PONTO DE
SOLDA
25
15
DETALHE DA FIXAÇÃO
REGISTRO
TIPO GAVETA

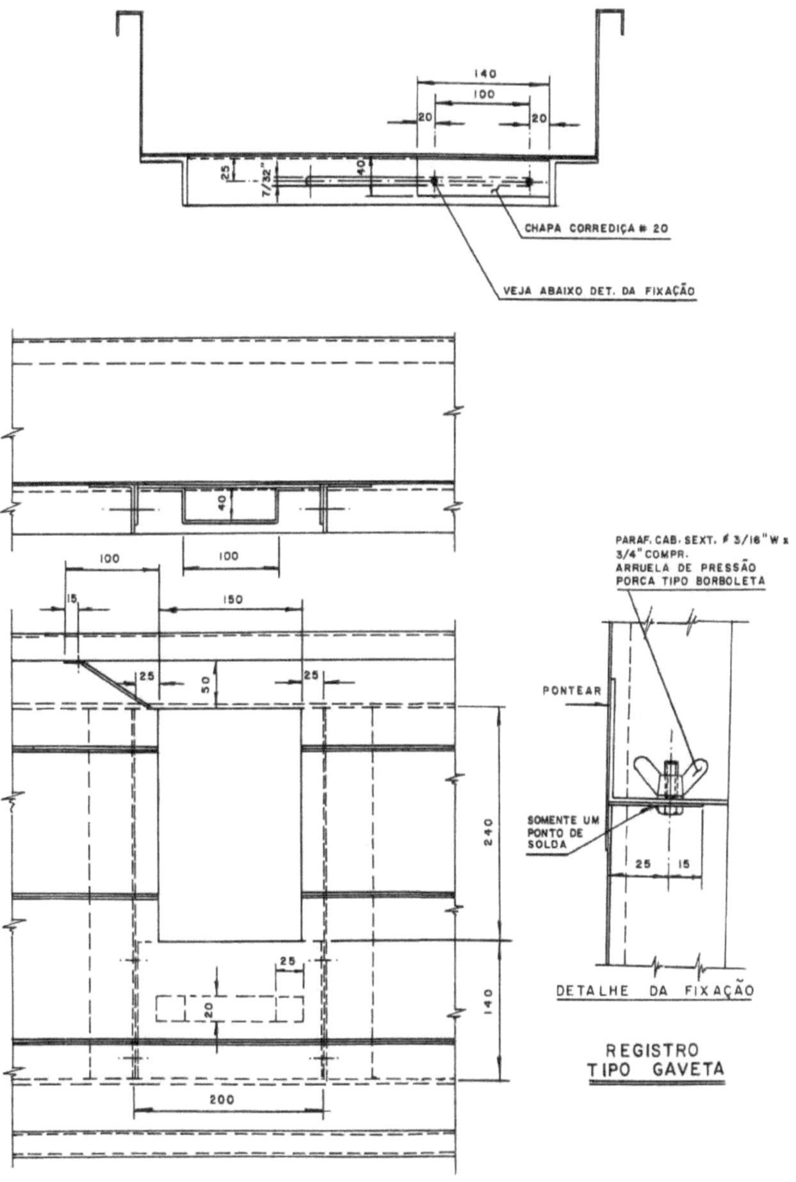

FIGURE 21

TAPETE DE FUMO
FLUXO

ERRADO
(INTERFERÊNCIA DOS CLEATS)

ERRADO
(INTERFERÊNCIA DOS CLEATS)

ESPAÇO LIVRE
CERTO
(IDEAL)

100
(MÍNIMO)
CERTO
(RECOMENDADO)

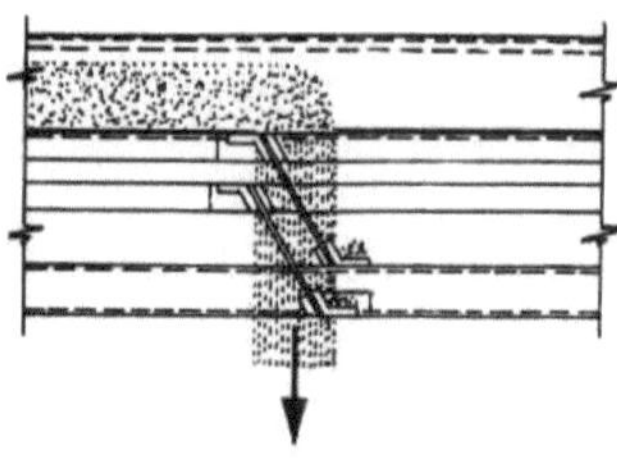

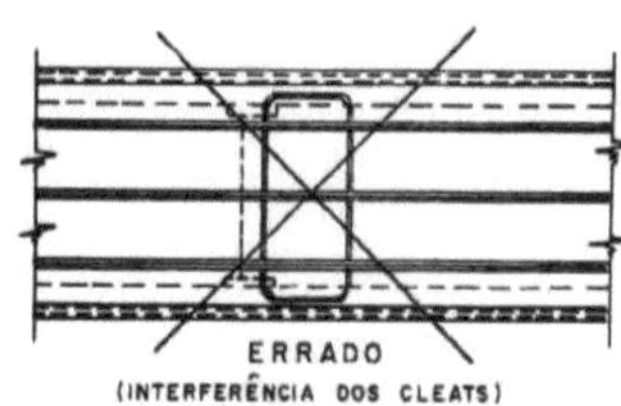

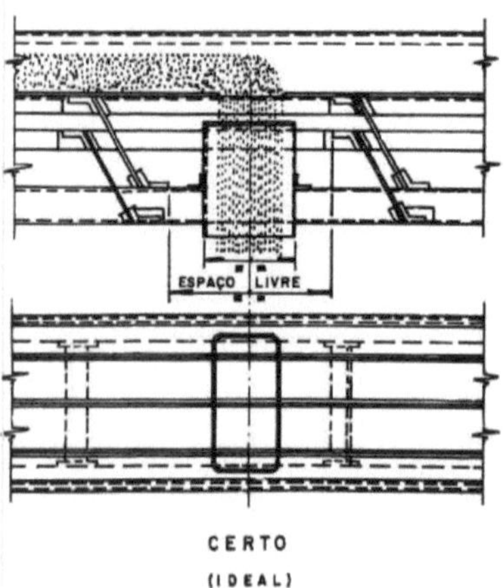

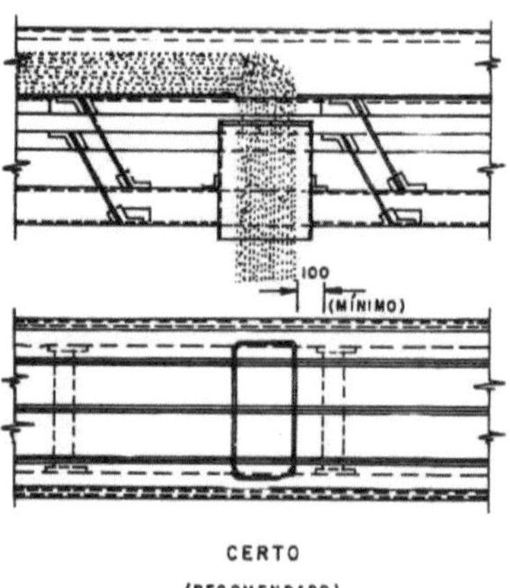

20

FIGURE 22

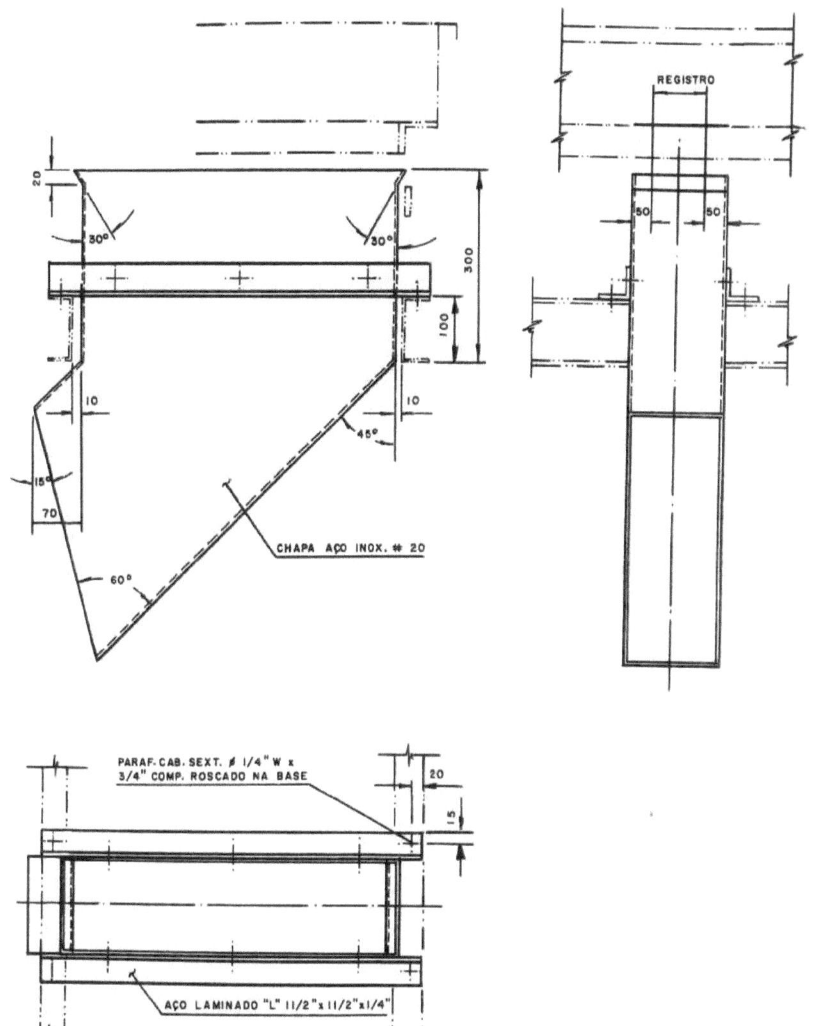

EXAMPLE OF A SMOKE CHUTE WITH LATERAL DIVERSION

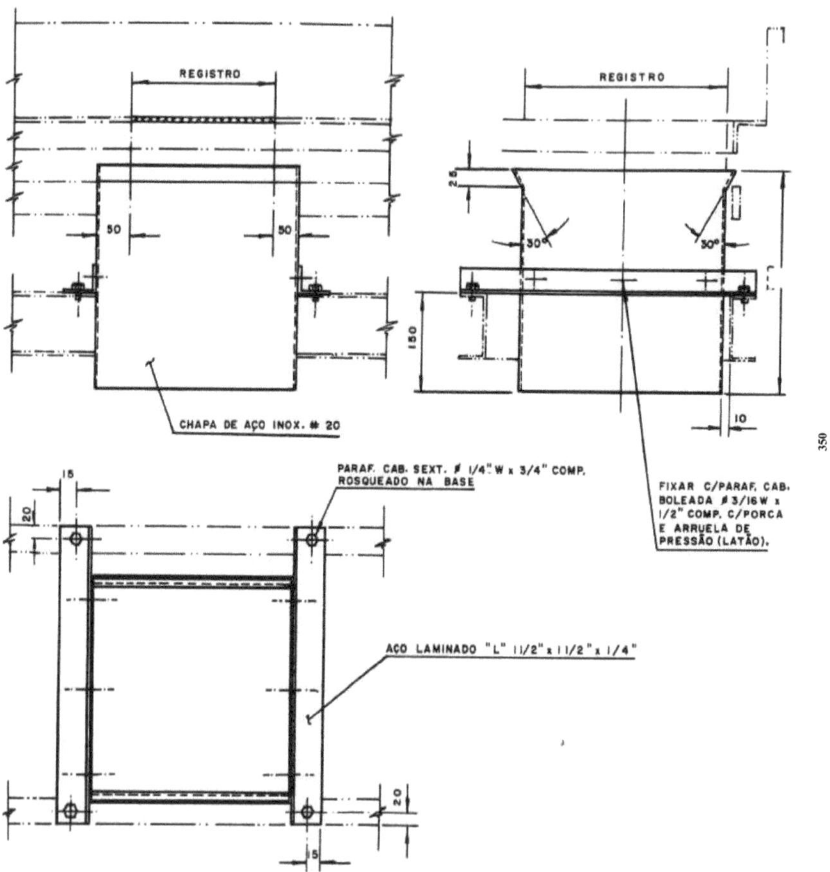

STRAIGHT SMOKE CHUTE

3.2 - CHASSIS

The rail is attached to the chassis. Figures 24 and 25 show the chassis and its components.

The chassis components are:

- 2 aluminium angles 2" x 2" x 1/4" to the length of the rail. The trunking is fixed to these brackets as shown in figure 8;
- The number of cleats on the chassis is variable and depends on the design of the conveyor. The spacing between the cleats is the same. The cleats are fixed to the aluminium angles using four 3/8" W x 1^" diam. bolts with nuts and lock washers.
- Figure 25 shows the details of how the springs are attached to the cleats. Each

cleat can be fitted with 2, 3, 4 or 5 springs, depending on the design of the conveyor.
- The cast iron chassis support cleats are welded to the base subassembly.

FIGURA 24

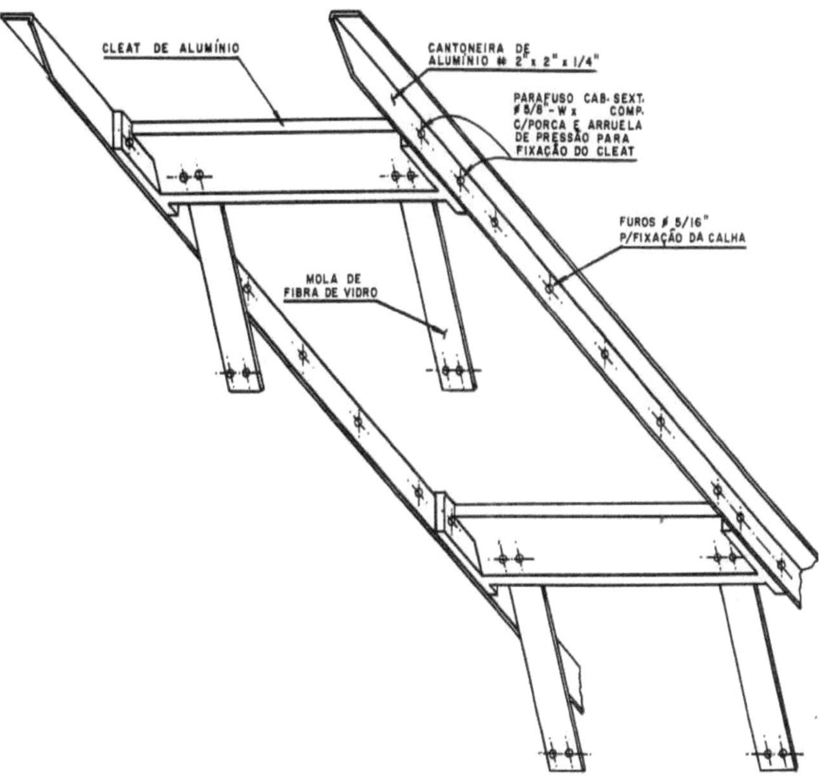

FIGURA 25

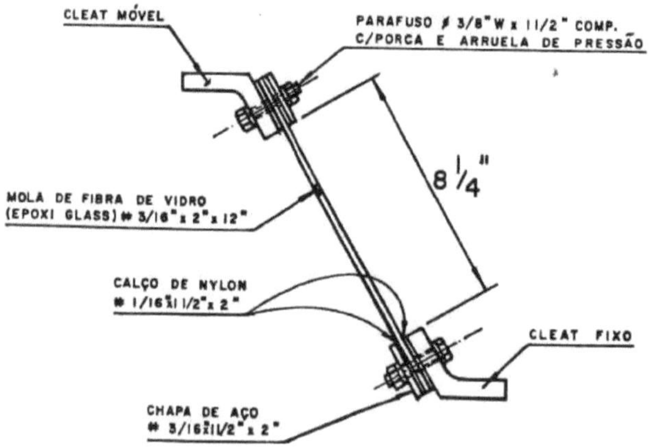

3.3 - BACKGROUND

Figure 26 shows the counterweight assembly. The components of the counterweight assembly are as follows:

- 2 rolled steel bars with a rectangular cross-section;

With a width of 2", the thickness of which is determined in the conveyor design. The counterweight bar must be at least flush with the counterweight ear and at most 200mm (overhang) from the centre line of the cleat hole, as shown in figure 26.

- Cast iron cleats;

The number of cast iron cleats is the same as the number of aluminium cleats in the chassis subassembly and even the spacing between them is the same. The cast iron cleats are fixed to the laminated steel bars using four 3/8" diam. bolts with nuts and washers. The length of the bolts varies depending on the thickness of the steel bar.

- Fibreglass springs;

The number of springs in the counterweight is equal to the number of springs in the chassis sub-assembly.

- Cast iron cleats;

These cleats are fixed to the base subassembly using 4 x 3/8" x 1^" diam bolts with nuts and washers.

FIGURE 26

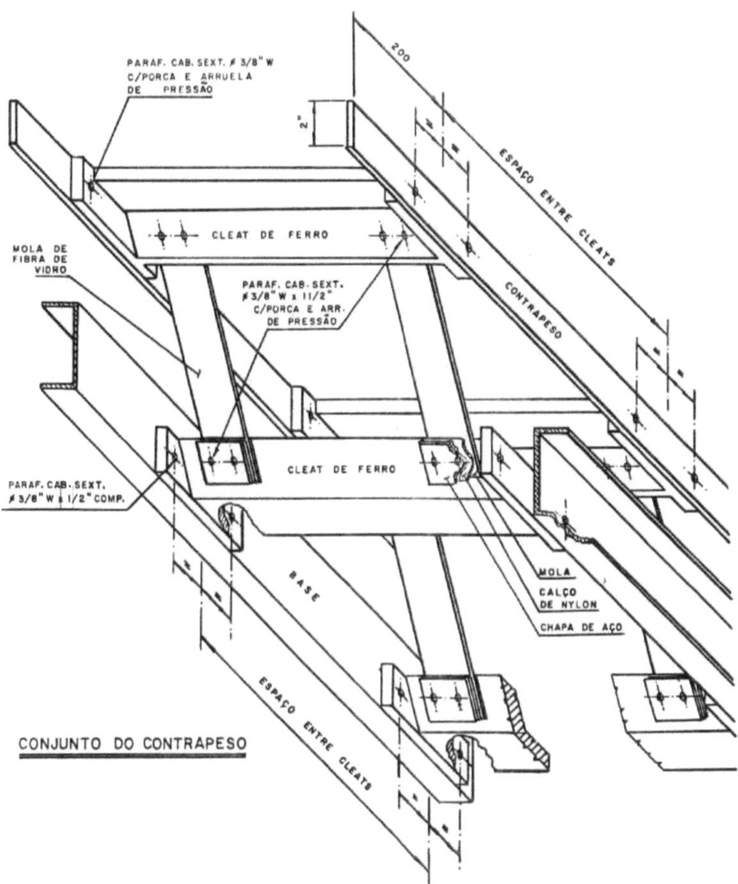

25.4- BASE

The base is rectangular in shape, made from "U" iron profiles measuring 4" x 1 5/8" x 1/4" with flanges facing outwards. The cast iron cleats are welded to the chassis and the cast iron cleats of the counterweight are fixed to the base using 4 bolts measuring $3/8$" x $1^1 Z>$", with nuts and washers, as shown in figure 27. The conveyor drive can be mounted at an intermediate point on the base or at the end, as shown in figures 28 and 29. The width of the base must be 2 mm longer than the length of the counterweight cleats as shown in figure 30.

FIGURE 27

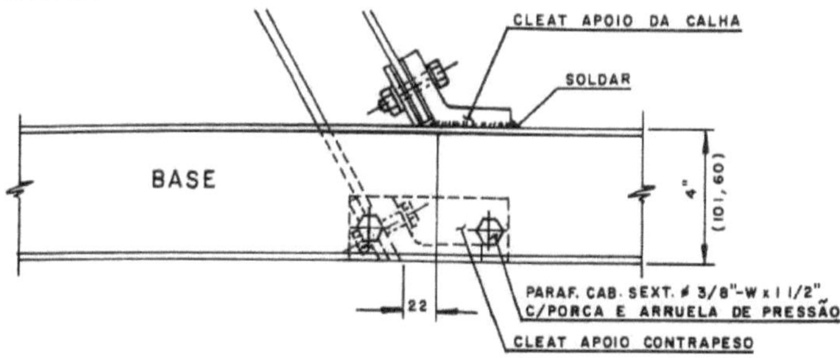

FIGURE 28

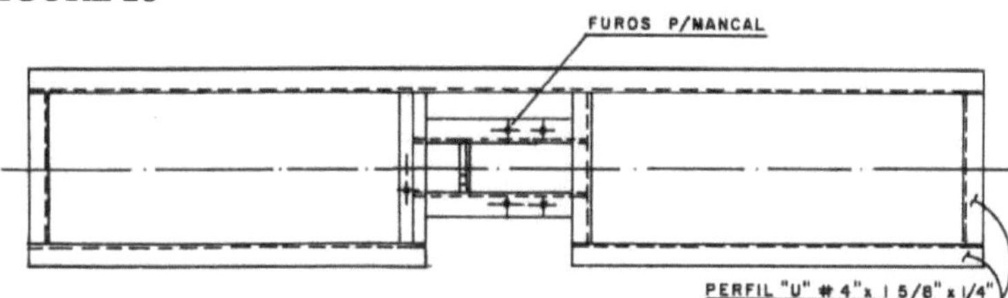

FIGURE 29

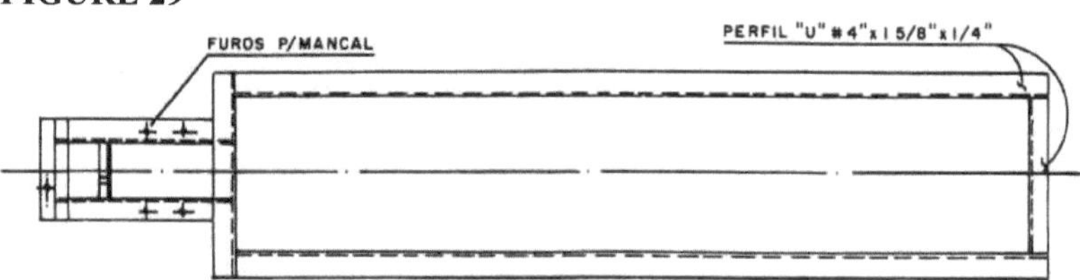

FIGURE 30

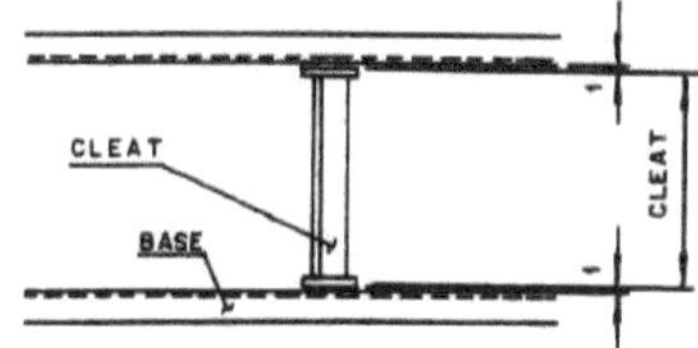

3.5 - DRIVE

The drive transmits movement to the chassis and counterweight sub-assemblies with the following components:

- Electric motor base;
- Electric motor;
- V-shaped driving and driven pulleys;
- Double eccentric shaft;
- 2 connecting rods;
- 2 adjusting screws with special nuts;

- 2 fibreglass springs;
- 1 aluminium cleats to drive the chassis;
- 1 steel cleats for the counterweight drive;
- 2 x 1" diam. bearings;
- 2 spring supports;
- 4 nylon shims;
- 4 steel shims;
- Stretcher.

There are 3 different types of drive for mounting on conveyors:
Type I - Drive located in the middle of the conveyor as shown in figure 31. In this case, the distance between cleats must be greater than or equal to 750 mm; Type II - Same as type I, but with the motor between the counterweights. This type can only be used on conveyors with a distance between cleats equal to or greater than 1000 mm, as shown in figure 32;

Type III - Drive at the end of the conveyor, as shown in figure 33.

FIGURE 31

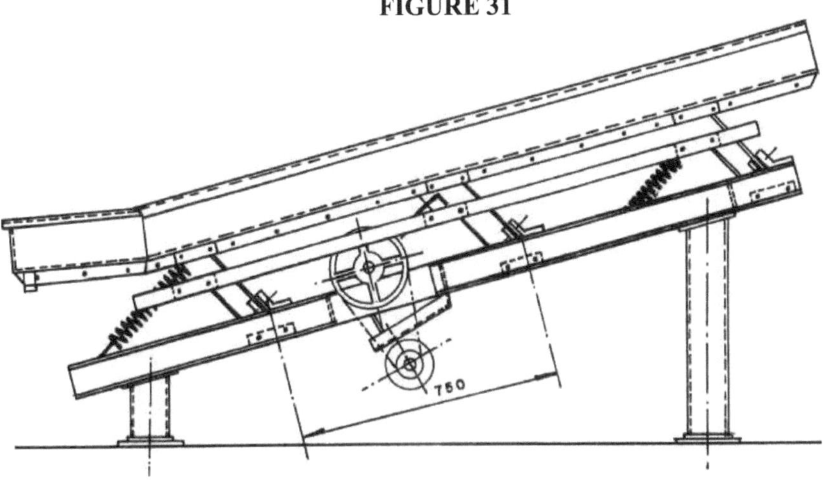

FIGURE 32

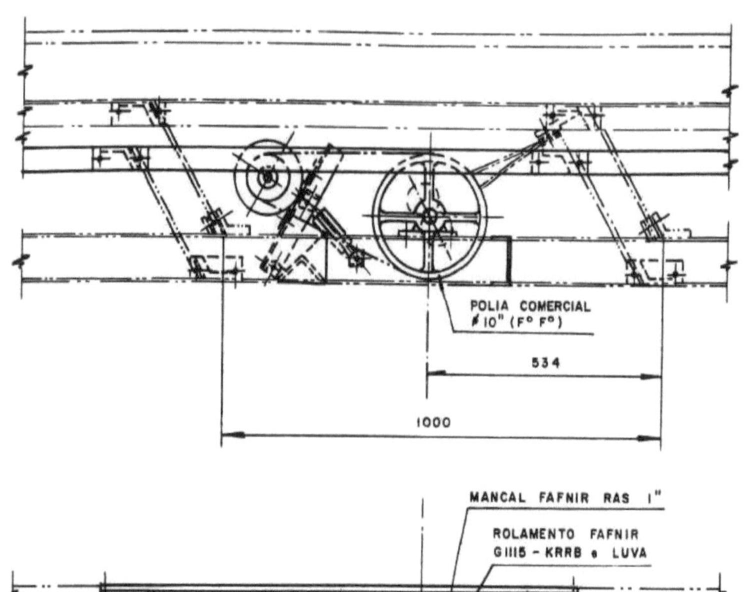

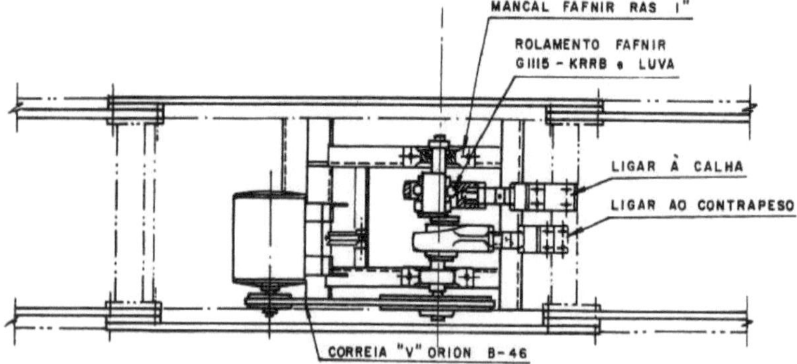

DRIVE ASSEMBLY AND COMPONENTS

FIGURE 33

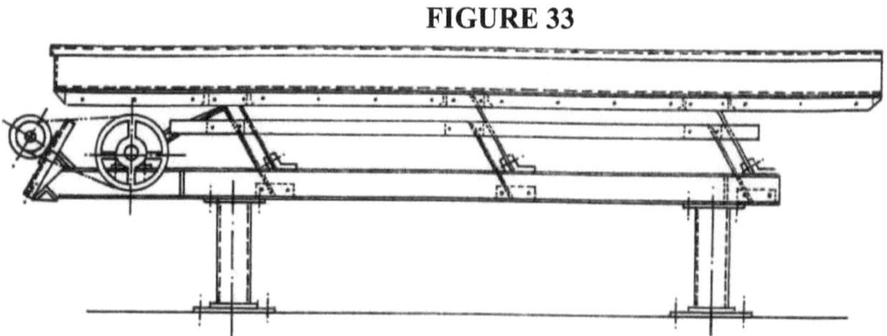

3.6 - COMPENSATION SPRINGS

When we tilt a conveyor without the drive springs, the springs in the chassis and counterweight deform, changing the natural frequency.

This deformation becomes greater the greater the inclination, length and width of the conveyor.

In order to return the conveyor to its original position and maintain the natural frequency, compensation springs are installed in the chassis and counterweight assembly, as shown in Figures 34 and 35. We suggest using compensation springs in vibrating conveyors with an inclination greater than 10° and a length of 6 metres.

3.7 - LONG CONVEYOR DIVISION

To allow for transport on a normal lorry, as well as mobility within factories for installation, the conveyors are divided into parts as shown in figure 36.

FIGURE 34

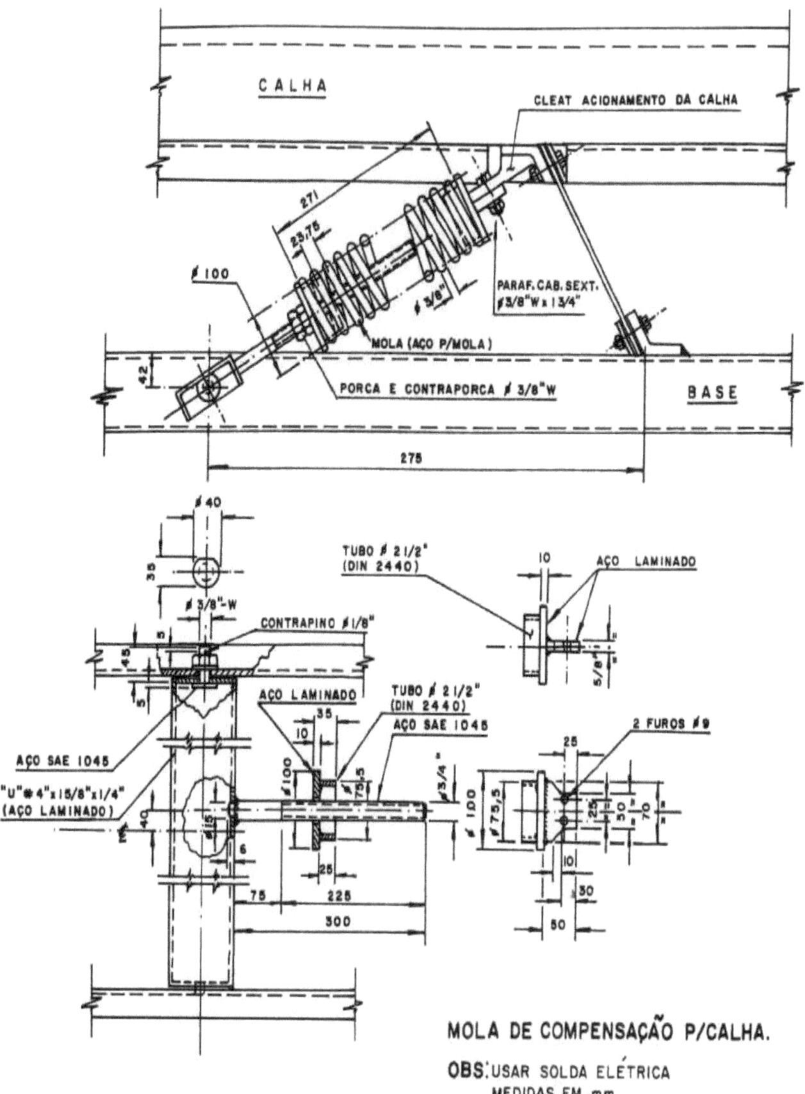

MOLA DE COMPENSAÇÃO P/CALHA.

OBS: USAR SOLDA ELÉTRICA
MEDIDAS EM mm.

FIGURE 35

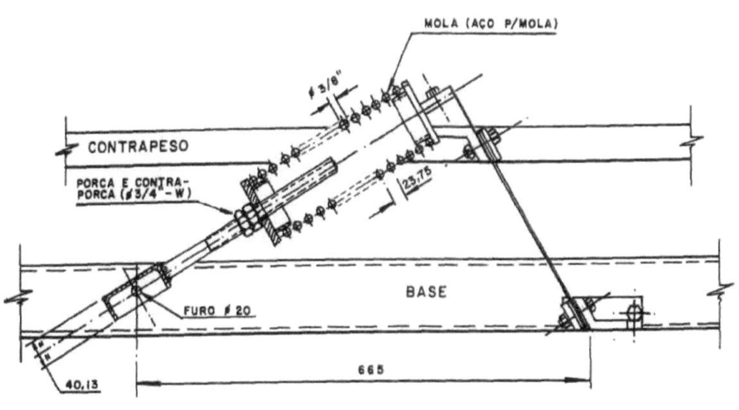

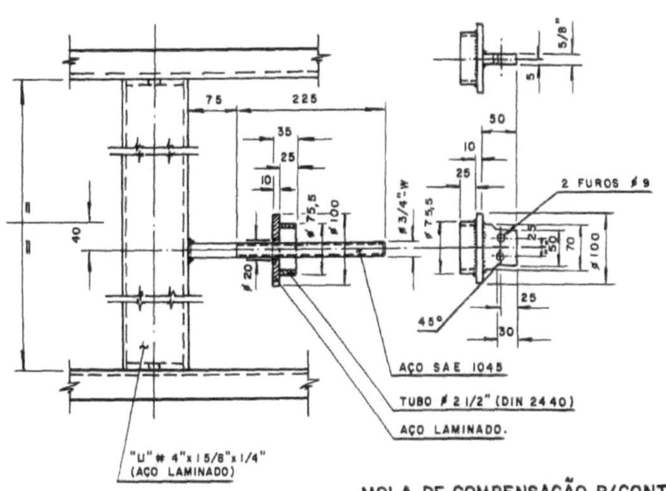

MOLA DE COMPENSAÇÃO P/CONTRAPESO.

OBS: USAR SOLDA ELÉTRICA
MEDIDAS EM mm.

31

FIGURE 36

300
CALHA
CHASSI
450
CONTRAPESO
250
BASE
250
30
25
70
25
300
350
100
LIGAÇÃO DA CALHA
50
100
10
15
15
LIGAÇÃO DO CHASSI
30
10
LIGAÇÃO DO CONTRAPESO
35
25
150
50
EMENDAS DO
TRANSPORTADOR
20
LIGAÇÃO DA BASE

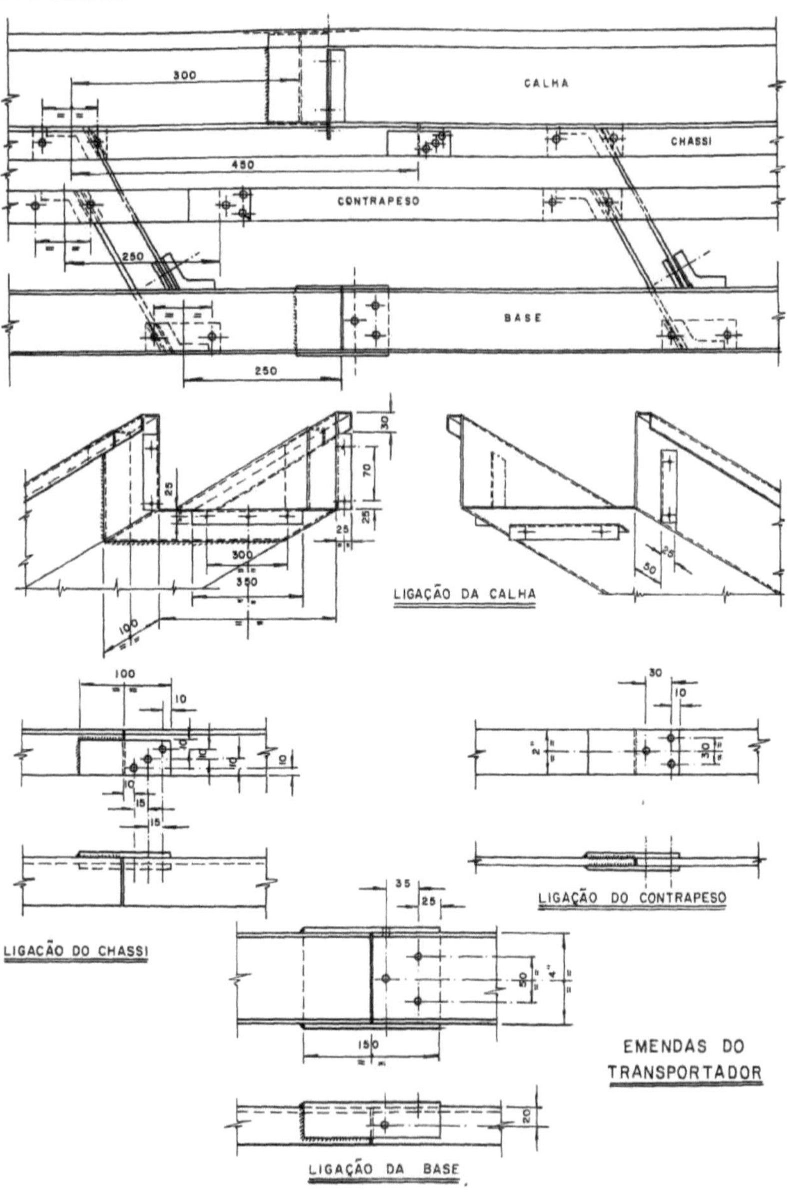

CHAPTER 4

FLOW CALCULATION

The flow rate of vibrating conveyors is calculated using the formula below:

$$Q = A \times B \times P \times V, \text{ where:}$$

Q = Flow in kg/h;
A = Gutter width in metres;
B = Height of the product mat in metres;
P = Specific weight of the product in kg/m3 (in the vibrating chute);
V = Product velocity in the chute in m/h;

The specific weight of the product when the chute is vibrating is determined experimentally and depends on the characteristics of the product itself and the rotation of the eccentric. Graphs Nos. 2 to 14 were developed by Imperial Tobacco Co. of Canada Ltd. and show the specific weight of various types of tobacco (with their moisture content) as a function of the rotation of the eccentric.

The specific weight shown in these graphs is for reference only, but should be measured whenever possible. This observation applies to different types of smoke or products used in the food, chemical, pharmaceutical and other industries.

NOTE: In the situation shown in the figure above, you should always have C equal to or as close as possible to 50 mm.

As you increase the eccentric's speed above 480 rpm, the conveyor's noise level rises; in addition, during product transport there is a very sharp separation between the lighter and heavier particles. The lighter particles take up the upper part of the product belt and the heavier ones the lower part. This makes the mixture heterogeneous and can, in some cases, cause production problems.

This aspect must be analysed very carefully, especially from the point of view of where the conveyor will be used on the production line. For example, when transporting tobacco leaves and/or stalks, the 430 to 550 rpm range can be used without harming the process; however, the "noise" aspect must not be forgotten.

Below is an example of calculating the flow of smoke in a vibrating conveyor.

Example:
Calculate the maximum flow rate of a vibrating conveyor with a chute width of 760 mm and a height of 150 mm, an inclination of 10° to the horizontal and an eccentric rotation of 450 rpm, for transporting threshed tobacco with a moisture content of 18 to 20%.

From the formula seen above:

$$Q = A \times B \times P \times V \text{ We will have:}$$

Q = flow rate in kg/h;
$A = 0.760$ m;
$B = 0.100$ m ($150 - 50 = 100$ mm);
$P = 44.50$ kg/m3, according to Graph 4;
$V = 795$ m/h, as shown in Graph 7;
$Q = 0.760 \times 0.100 \times 44.5 \times 795 = 2688$ kg/h

Moisture conversion (%)
By way of illustration, below is an example of transforming the calculation of a flow rate with a certain humidity into one with a different humidity:

Let's say we have a flow rate of 7872 kg/h at 30% CRS humidity; if we need to calculate its new value from this flow rate, but at 13%, we should proceed as follows: Q at 30% = 7872 kg/h Q at 13% = 7872 x (100 - 30) / 100 - 13 = 7872 x 70/87 Q at 13% = 6333 kg/h

DRY CRS OR STALK   12 % HUMIDITY *SHAFT ECCENTRICITY7/6"*

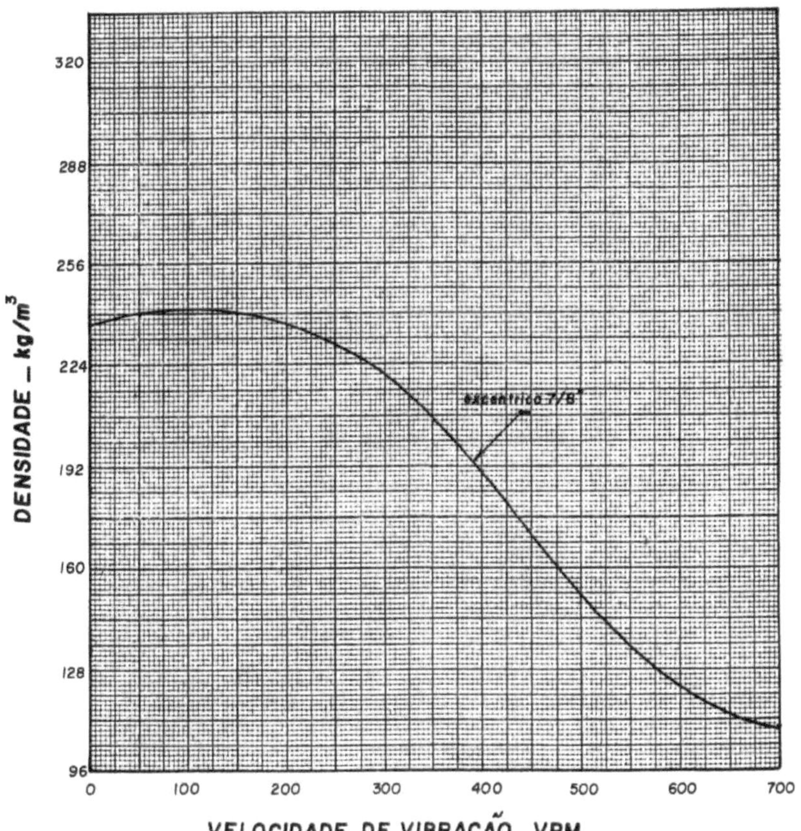

CUTTING SMOKE   *10% ADE OF* SHAFT ECCENTRICITY: 7/8"

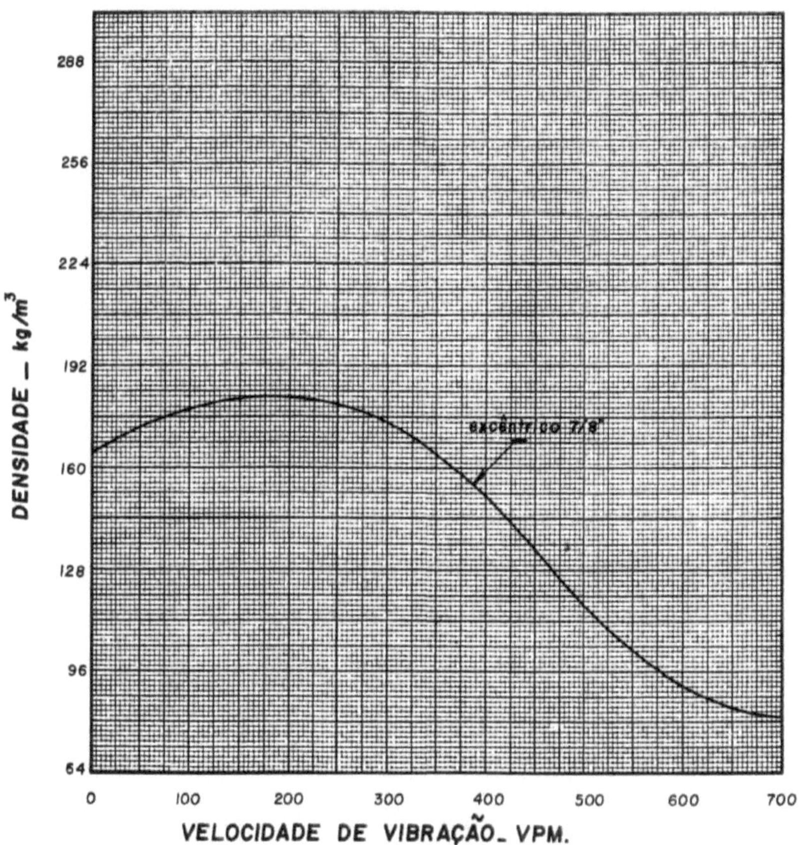

36

GRAPH NM

DEBULKED SMOKE - 19% OF AN AGE. SHAFT ECCENTRICITY: 7/8"

256
224
192
160
128
96
64
32
DENSIDADE — kg/m³
excentrico 7/8"
0 100 200 300 400 500 600 700
VELOCIDADE DE VIBRAÇÃO _ VPM.

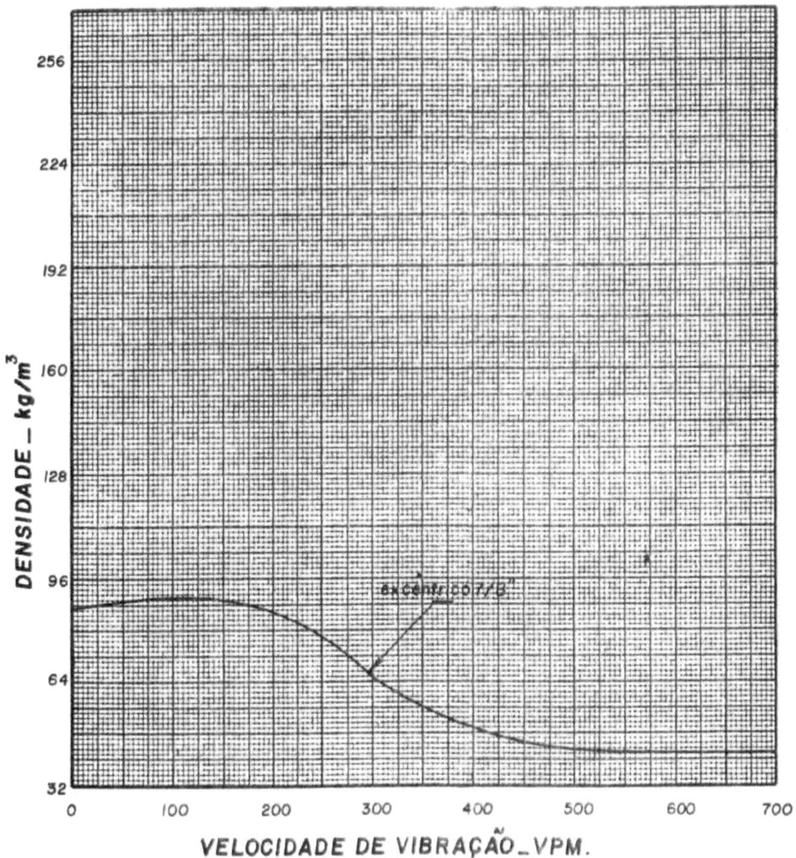

CUTTING SMOKE   14 % *HUMIDITY* (55 cuts per inch) *EXCENTRICITY DQ EIXQ'. 7/8"*

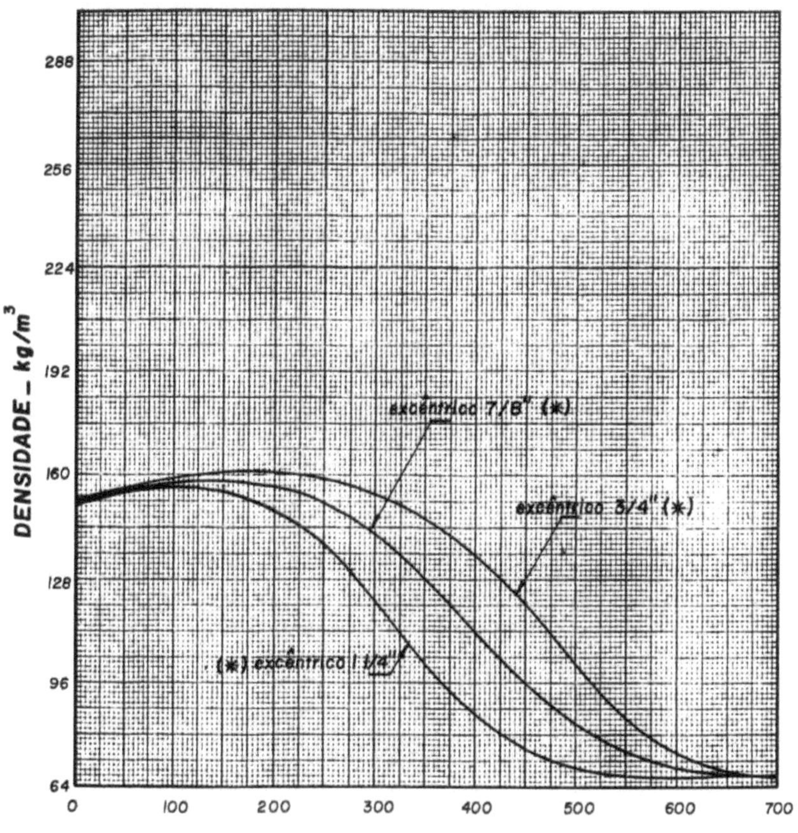

CRS _30% HUMIDITY OR <u>WTS.</u>

<u>AXLE ECCENTRICITY</u>; 7/8"

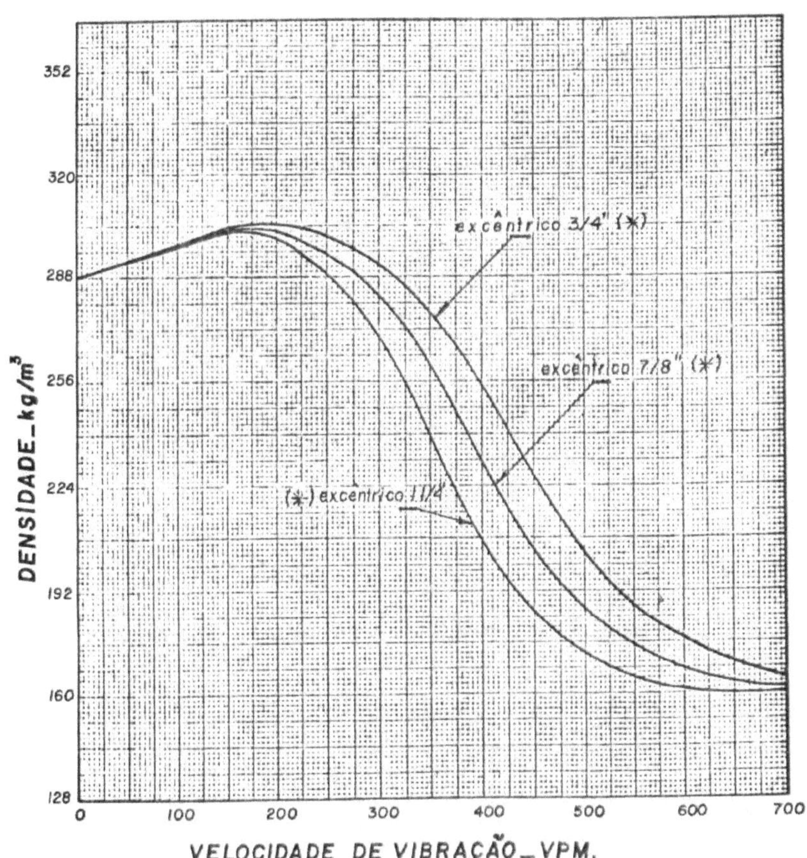

(✳) Normalmente usa_se o excêntrico de 7/8".

DEBULHED SMOKING (veneers). 15% or 20% ONE AGE.
VIBRATION AMPLITUDEb(eccentric):7/8"

CHART N 7?

1736,60
1645,20
1553,80
1462,40
1371,00
1279,60
1188,20
1096,80
1005,40
914,00
822,60
731,20
639,80
548,40
457,00
365,60
VELOCIDADE DE TRANSPORTE — m/h
VELOCIDADE DE VIBRAÇÃO_ VPM.
100
200
300
400
500
600
700
0°
5°
10°
15°
20°
INCLINAÇÃO
DA
CALHA

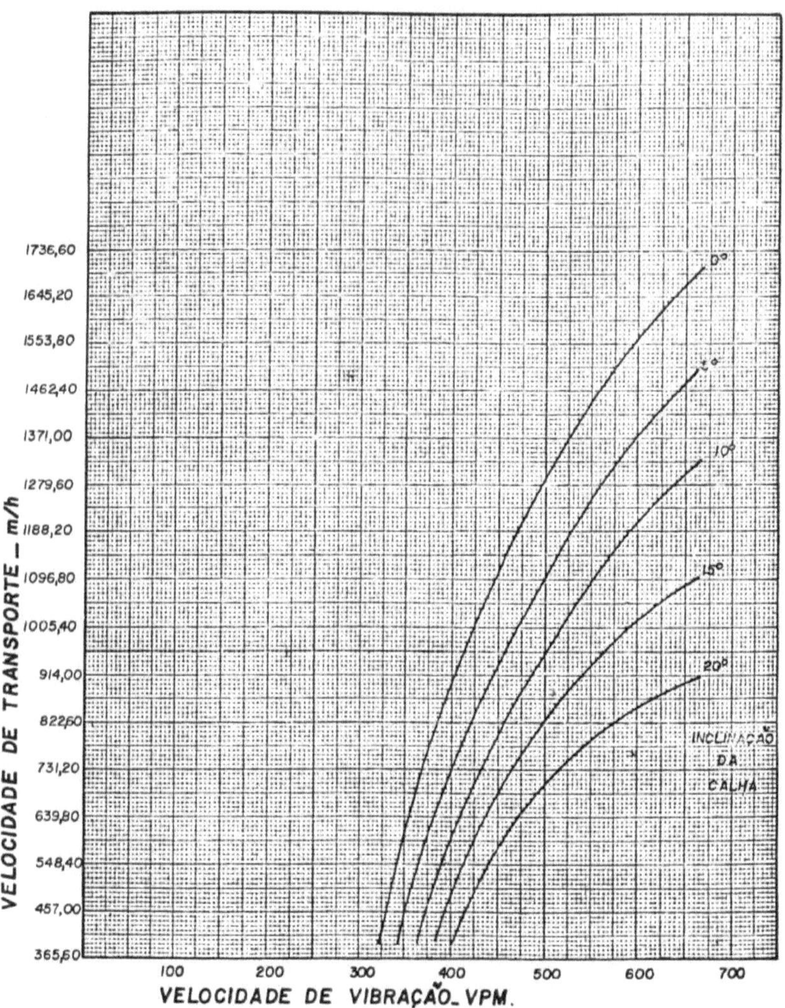

40

CHART N 8ᵖ

CUT SMOKE WITHOUT CRS (55 cuts per inch). 187o to 207o HUMIDITY.

VIBRATION AMPLITUDEQ (eccentric) (7/8")

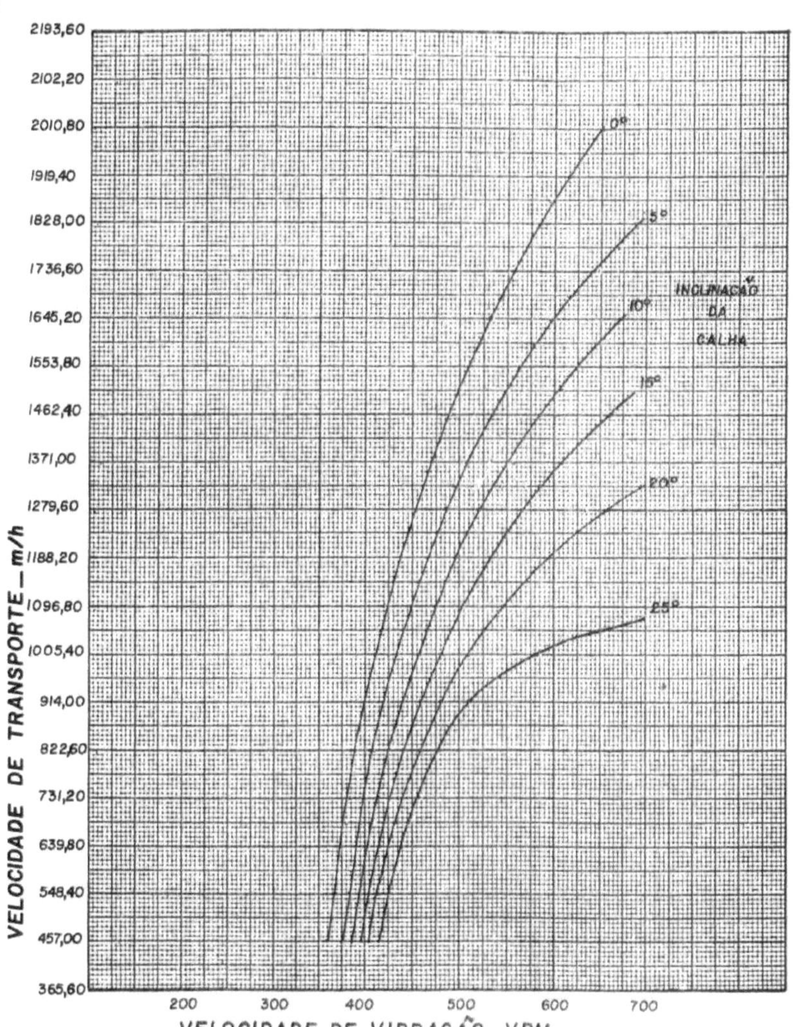

41

CUTTING SMOKE - 137 _to_ 157 _ONE AGE._
VIBRATION AMPLITUDEQ (eccentric). '7/8 "

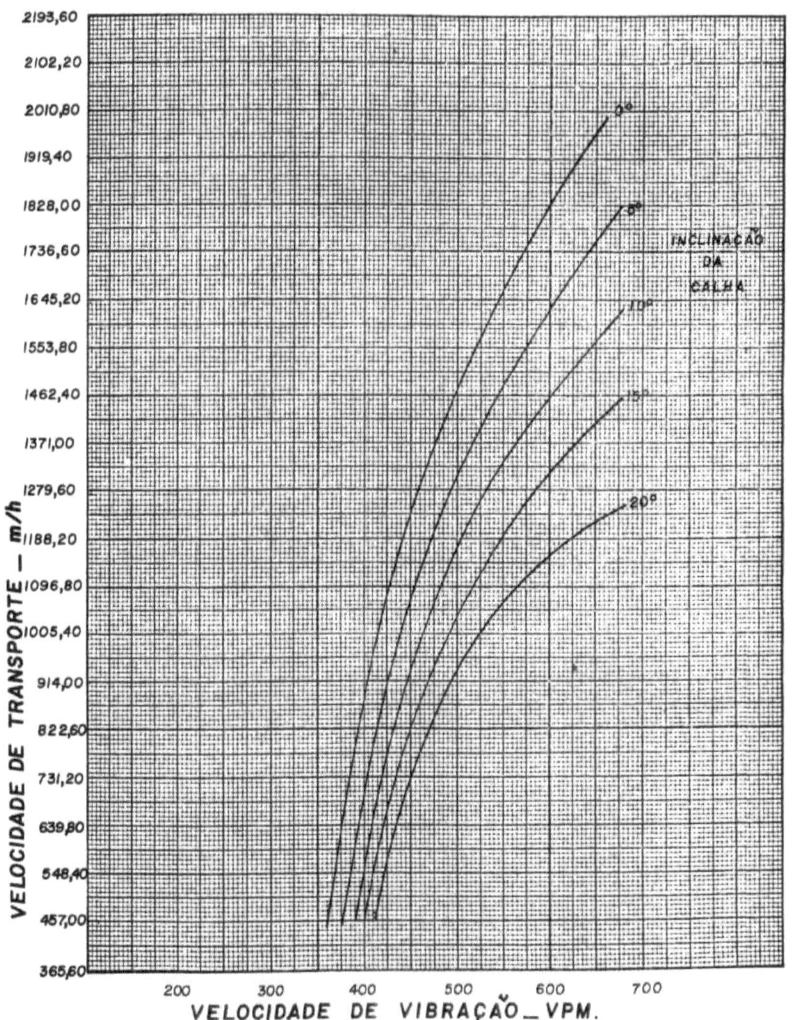

CHART Nᴾ 10

CRS, 30% *UM/DADE OR* <u>*WTS*</u>.

<u>*VIBRATION AMPLITUDEQ (eccentric):* 7/8</u>

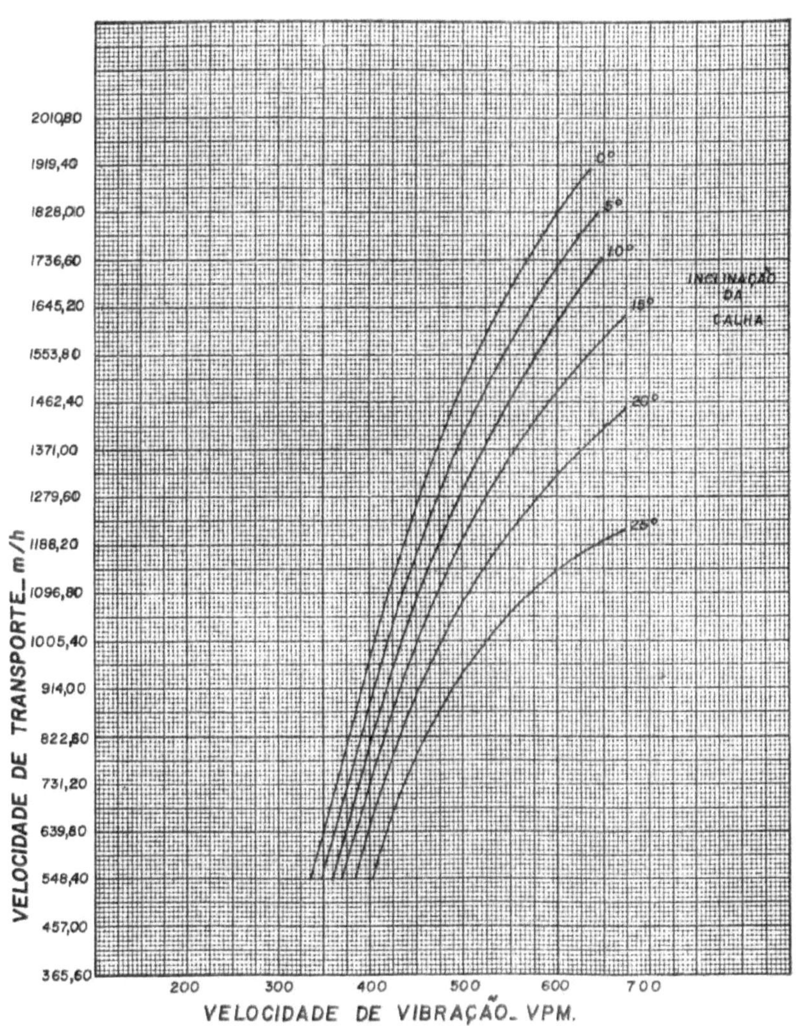

43

DRY CRS OR STALK. 137o MOISTURE VIBRATION AMPLITUDEQ(eccentric)7/8'

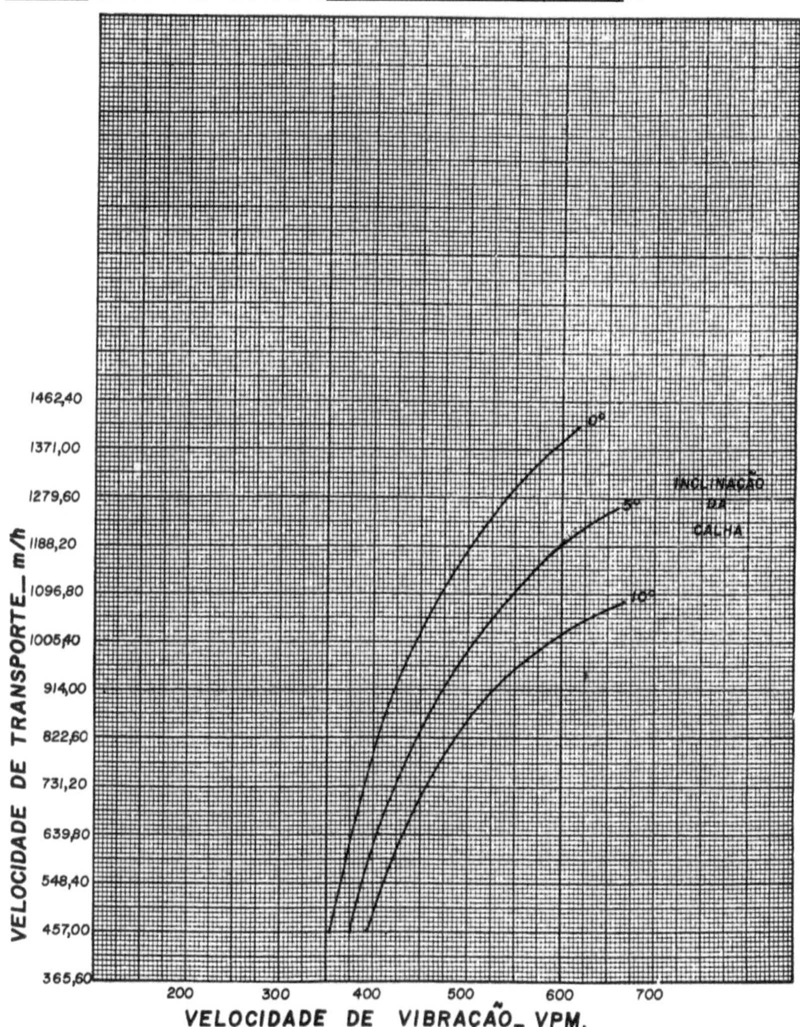

CHART Nº 12

THRESHED LAMINA SMOKE-14% AN AGE.

VIBRATION AMPLITUDE (eccentric) 7/8"

VELOCIDADE DE TRANSPORTE — m/h.
1919,40
1828,00
1736,60
1645,20
1553,80
1462,40
1371,00
1279,60
1188,20
1096,80
1005,40
914,00
822,60
731,20
639,80
548,40
457,00
365,60
200
300
400
500
600
700
VELOCIDADE DE VIBRAÇÃO. VPM.
0º
5º
10º
15º
20º
INCLINAÇÃO DA CALHA

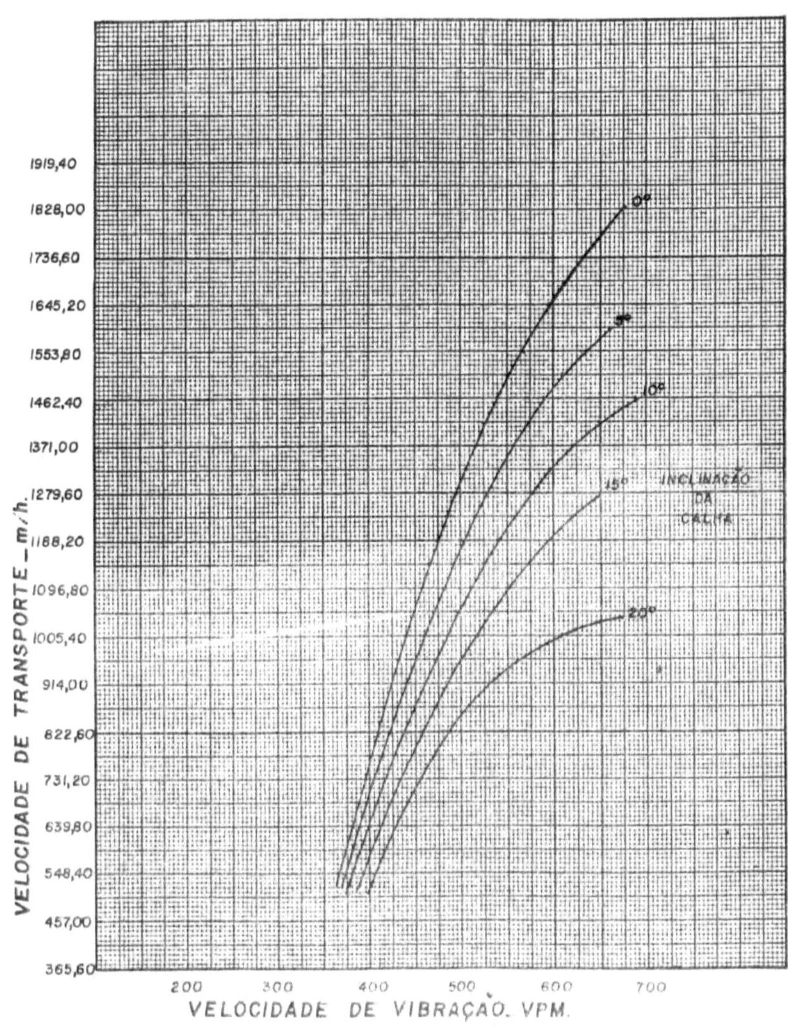

LAMINATED TALK OR CRS OF THIS- 197th to 207th AN AGE.

VIBRATION AMPLITUDE (eccentric): 7/8"

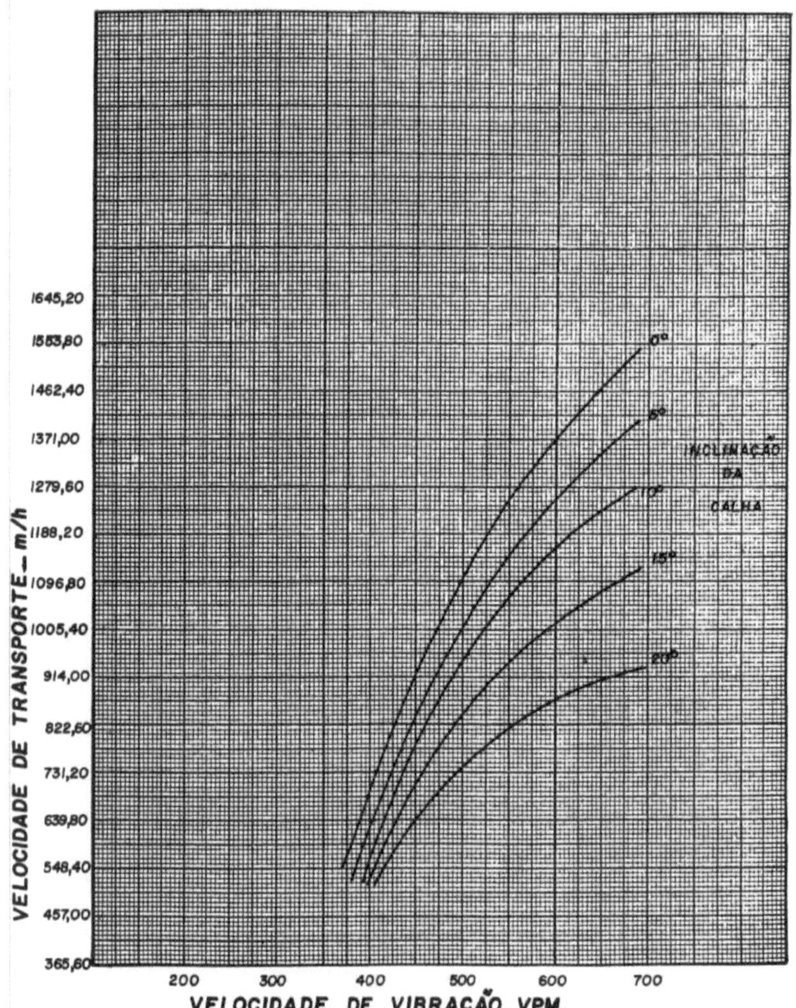

CHART Nº 14

LAMINATED TALUS OR CRUSHED STEEL - 30% HUMIDITY.
VIBRATION AMPLITUDEQ (eccentric).'7/8"

0°
5°
10°
15°
20°
INCLINAÇÃO
DA CALHA
1553,80
1462,40
1371,00
1279,60
1188,20
1096,80
1005,40
914,00
822,60
731,20
639,80
548,40
457,00
365,40
VELOCIDADE DE TRANSPORTE — m/h
200
300
400
500
600
700
VELOCIDADE DE VIBRAÇÃO— VPM.

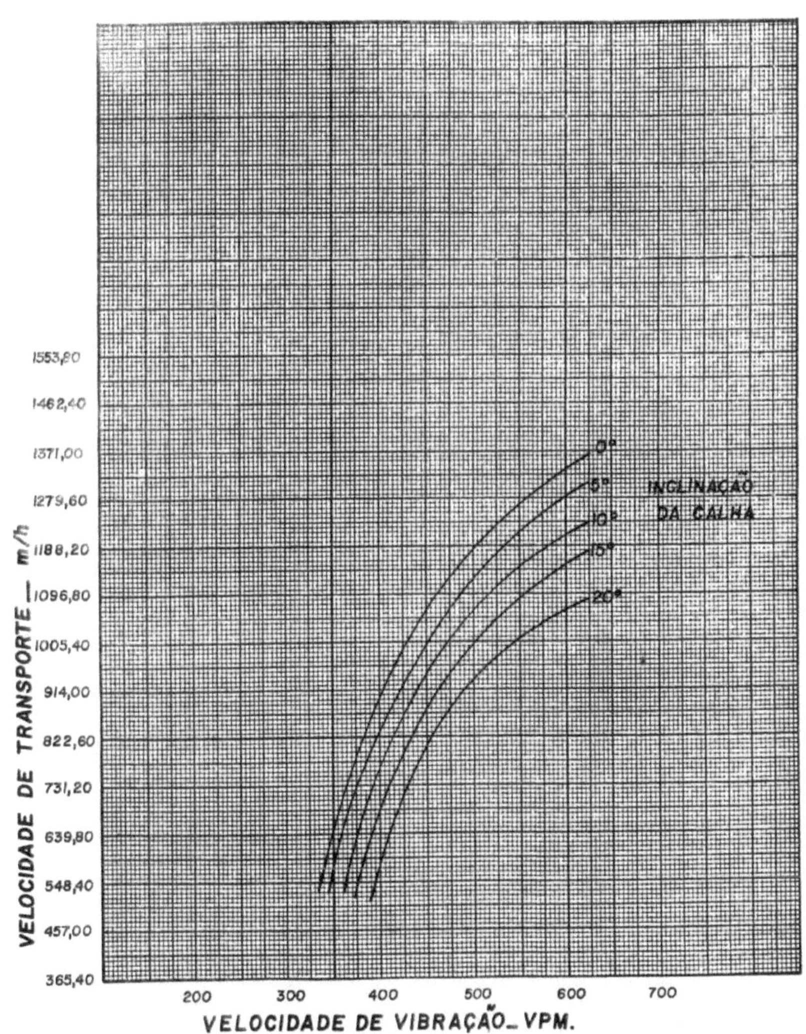

CHAPTER 5

PROJECT

Below is a list of the essential points that must be taken into account when carrying out a balanced vibratory conveyor project.

5.1 - PROJECT DATA

The following data is required to run the project:
- Type of product and its humidity;
- Product flow;
- Inclination of the conveyor with the horizontal;
- Length and balance;
- Details of registers, internal divisions or others.

5.2 - RAIL SIZE AND ECCENTRIC ROTATION

Determining the flume size and eccentric rotation to meet the design flow is done by trial runs calculating the flow for various standardised flume sizes and eccentric rotations.

For ease and from experience, we should start with the following value: set the height of the chute at 150 mm. From this value, calculate the flow rate of the vibrating conveyor at standard widths from 480 mm onwards, until the desired flow rate is reached.

As a safety margin, we recommend designing the conveyor for a flow rate 30 per cent higher than the specified flow rate, in order to meet temporary production increases or those that may become permanent due to plant expansion. It should also be borne in mind that due to spring fatigue, it is recommended that the eccentric speed be calculated at 5% higher than its natural frequency (VPM).

5.3 - SPRING WEIGHT CALCULATION

Once we know the rotation of the eccentric, we enter this value in graph 1 and determine the weight that each spring will be subjected to. In the experience of the Imperial Tobacco Co. of Canada Ltd., the maximum permissible weight per spring should not exceed 8kg.

5.4 - CALCULATING THE DISTANCE BETWEEN CLEATS

Knowing the width and height of the chute, as well as the weight each spring is subjected to, we can determine the distance between the conveyor cleats. In this case, we have 4 alternatives:

- 2 springs per cleat
- 3 springs per cleat
- 4 springs per cleat
- 5 springs per cleat

The distance "X" between cleats, as shown in the figure below, is calculated using the formula below:

$$X = (M.N. - Q) / L$$

X = Distance between 2 underlying cleats in metres;
M = Weight a cleat is subjected to in kilos;
N = Number of elastic springs in 1 cleat;
Q = weight of the complete cleat (with bolts, nuts, etc.) in kilos;
L = Weight per linear metre of the complete gutter without cleats in kilos;

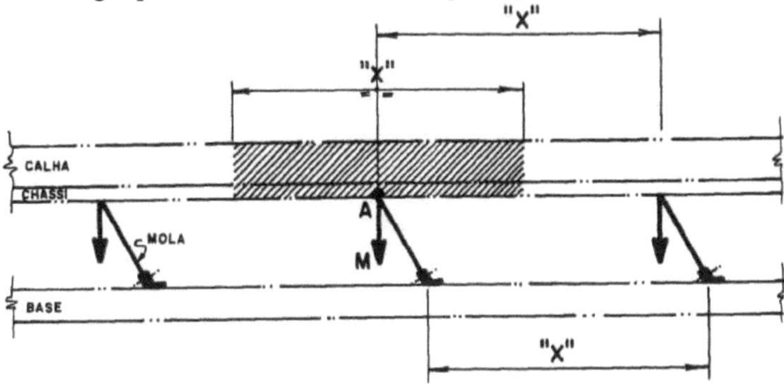

The distance between cleats "X" must not exceed 1,200 mm according to Imperial Tobacco Co. of Canada Ltd. On conveyors with registers, it is often necessary to start with the distance between cleats already established, so as not to cause the product discharge to interfere with the distance between cleats, as shown in figure 21. Once the distance between cleats has been set, the reverse calculation is made to determine the values of the other conveyor elements.

In order to calculate the values of L and Q, we present below a list of the weights of the components of vibratory conveyors:

- Stainless steel sheet no. 20 = 8.09 kg/m2;

- Aluminium angle bracket 2" x 2" x 1/4" =1.64 kg/m;

- Weight of single and drive aluminium cleats:

Transp. width (mm)	Simple (kg)	Drive (kg)
1040	3,555	4,047
900	3,088	3,580
760	2,621	3,113
620	2,500	3,200
480	1,900	2,500

- Total weight of the cleat spring fastener assembly = 0.150 kg; Comprises the following parts:

- 2 x 3/8" W x 1 ⅜" diam. screws with nut and spring washer = 2 x 0.0375 kg; - 2 x 1 7₂ " x 1/16" nylon shims = 2 x 0.004 kg;
- 1 steel plate 2" x 2 7₂ " x 3/16" = 1 x 0.070 kg;

• Total weight of the aluminium cleat fixing set = 0.150 kg; Includes: 4 screws diam. 3/8" W x 1 ^" with nut and spring washer;

• Weight of 1 1/4" x 5/8" diameter bolt with nut and spring washer for fixing the rail to the chassis = 0.020 kg;

• Weight of 1 1/4" x 5/8" diameter screw fixing the rail to the aluminium cleats = 0.015 kg.

5.5 - CALCULATING THE COUNTERWEIGHT BARS

It is a basic requirement that the weight to which the rail springs are subjected is equal to the weight to which the counterweight springs are subjected. This equality determines the total weight of the rectangular counterweight bars.

The weight components to which the rail springs are subjected are:

a) Complete rail with the following parts:
- U-shaped sleepers, as shown in fig. 5;
- 2" x 2" "L" shaped sheets, with screws, nuts and washers, as shown in fig. 8;
  - 15 x 50 mm "L" shaped sheets with 1/4" diam. screws, as per fig. 9;

b) Aluminium angle brackets;

c) Aluminium cleats with screws, nuts and washers for fixing to the aluminium corner;

d) Bolts, nuts, washers, nylon shims and steel plates for fixing the springs to the aluminium cleats.

The weight components that are exerting pressure on the counterweight springs are:

a) Cast iron cleats with screws, nuts and washers for fixing to the rectangular bases;

b) Rectangular bars;

c) Bolts, nuts, washers, nylon shims and steel plates for fixing the springs to the cast iron cleats;

d) To determine the counterweight bars, a list of their component weights is shown below:

Transp. arg. (mm)	Simple (kg)	Drive (kg)
1.040	12,887	13,900
900	11,124	12,137
760	9,361	10,374
620	7,200	8,000
480	5,600	6,300

e) Weight of 2" wide steel bars:

Thickness (pol.)	Weight (kg/m)
1/8"	1,27
3/16"	1,90
5/16"	3,17
3/8"	3,80
7/16"	4,43
1/2"	5,06
9/16"	5,69
5/8"	6,33
3/4"	7,59
7/8"	8.86
1"	10,12

f) Cleat spring positioning

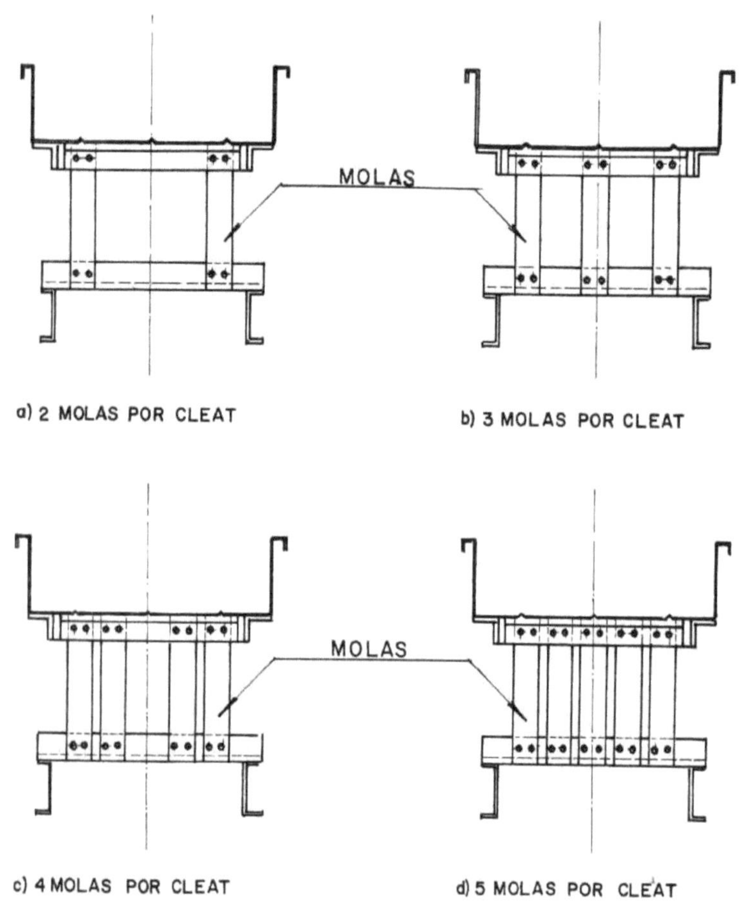

5.6 - DRIVE LOCATION

We saw in section 3.5 that we have 3 types of drive. Depending on the distance between cleats and interference that may arise in the installation, one of them must be chosen.

5.7 - ELECTRIC MOTOR SELECTION

The choice of electric motor is calculated using the formula below:

$$HP = rpm. \, N. \, P. \, t / 210,000. \, \% \, torque \quad where:$$

HP = Power of the electric motor in HP;
rpm = Rotations of the eccentric;

N = Number of springs supporting the rail;
t = Eccentricity in inches (standard = 7/8");
P = Force required to deflect a spring in t/2, which is 31 lbs for 3/16 springs and 67 lbs for 1/4" springs;
% torque = torque with the motor blocked, let's consider 150%;

The above formula was developed by the Imperial Tobacco Co. of Canada. Applying the data, we have:

HP (3/16" springs) = rpm. N. 31. 0,875/ 210.000 . 1,50
Considering the rpm of the eccentric is 500 rpm, we conclude:
HP (3/16" springs)) = N . 0.043
N (3/16" springs) = 23.25 x HP;

We can then establish the following table for 3/16" springs:

1 HP up to 23 springs;
1.5 HP from 24 to 25 springs;
2 HP from 26 springs to 46 springs;
3 HP from 47 to 69 springs;
5 HP from 70 to 115 springs;

Using the same sequence of reasoning for 1/4" springs, we have the following table:

1 HP up to 10 springs;
1.5 HP from 11 to 15 springs
2 HP from 16 to 20 springs
3 HP from 21 to 30 springs;
5 HP from 31 to 50 springs

With the use of compensation springs, inclined conveyors will require a greater demand from the motor. In these cases, we suggest leaving a gap of 10 per cent of power in relation to the chosen electric motor.

5.8 - MANUFACTURING DRAWINGS

Below is a list of the drawings required to manufacture balanced vibrating conveyors:

ItemDescription
01   -Drive assembly;
02   -Drive rod;
03   -Special SAE diam nut   , 1 5/8" x12   threads   (right);

04   -Special SAE diam nut   , 1 5/8" x12   threads   (left);
05 - Adjusting screw SAE diam, 1 5/8" x 12 threads;
06   -Steel plate   2" x 1 7_2 " x 1/16";
07   -Nylon shim   2" x 1 7_2 " x 1/16";
08 -Springs (drive support);
09   -Drive shaft;
10   -Spring support;
11   -Motor pulley
12 - Base for 1, 1 7>, and 3 HP electric motors;
13 -Support;
14 -Pino ;
15 -Support;
16 -Pino ;
17 -Extender   ;
18   -Extension body;
19 -Cleat   for 480mm rail;
20 -Cleat   for 620mm rail;
21 -Cleat   for 760mm rail;
22 -Cleat   for 900mm rail;
23 - Cleat for 1040 mm rail;

5.9 - EXAMPLES OF CONVEYOR PROJECTS

The following is an example of the design of 2 cases of a vibrating conveyor:

- Case 1: Conventional horizontal conveyor;
- Case 2: Conveyor inclined horizontally.

5.9.1 - Case 1

Design 1 horizontal vibrating conveyor 6 metres long to transport 3,000 kg/h of stalks at 13% humidity.

5.9.1.1 - Choice of chute size (width and eccentric rpm); It is recommended that the design flow be 30 per cent greater than the predicted flow. So the design flow will be 3,900 kg/h.

From experience, we initially decided to use a 480 mm wide chute with a height of 150 mm (100 mm smoke mat). The eccentric speed was set at 450 rpm. The flow rate will then be calculated as follows:

$Q = A \times B \times P \times V$
$A = 0.480$ m (gutter width);
$B = 0.100$ m (height of the smoke mat);
$P = 168$ kg/m3 (specific weight of the material - graph 2);
$V = 1024$ m/h (transport speed - graph 11);
$Q = 0.480 \times 0.100 \times 168 \times 1024 = 8,257$ kg/h

The conveying capacity is well above the flow requirements of the project. So, as we are not using a chute smaller than 480 mm, we reduce the eccentric's speed to 430 rpm and we have:

V = 946 m/h   (graph 11);
P = 173.4 kg/m3 (graph 2);

The new design flow will be:
Q = 0.480 x 0.100 x 173.4 x 946 = 7873 kg/h

5.9.1.2 - Determining the weight to which each spring is subjected;

It is recommended that the rotation of the eccentric be 5% higher to take spring fatigue into account: 430 + 22 = 452 rpm. With this value, we enter graph 1 and find 2 values for the weight per spring corresponding to the 3/16" and 1/4" eccentric. As both are less than 8 kg per spring, we must analyse the economic aspect and the values we will find for the distances between cleats (X).

5.9.1.3 - Calculating the distance between cleats (X);

M. N = L . X + Q ; (See details in item 5.4)

M = Weight in kilos that a cleat is subjected to;
N = Number of elastic springs in 1 cleat;
L = Weight in kilos per linear metre of the complete chute without cleats;
X = Distance in metres between 2 underlying cleats;
Q = weight of the complete cleat (with screws, nuts, etc.);

We have the following alternatives to calculate:

M1 = 3.735 kg/spring for springs with a thickness of 3/16";
M2 = 7.695 kg/spring for springs with a thickness of 1/4";
N1 = 2 springs;
N2 = 3 springs;
N4 = 4 springs;
N5 = 5 springs;

a) Determination of Q (weight of a complete aluminium cleat):

- 1 aluminium cleat for 480 mm rail = 1,900 kg;
- Set attached to 1 spring on cleats = 0.150 kg;
- Plate for fixing the rail to the cleats, as per fig. 9 (calculate the area of the plate and multiply by the specific weight of stainless steel) = 0.157 kg;
- The weight of 1 1/4" x 5/8" screw for fixing the aluminium cleats to the chassis is

0.015 kg; as an average of 13 screws are used, the total weight is 0.195 kg;

We then have the following alternatives for the value of Q:

5 springs = 1.900 + (0.150 x 2) + 0.157 + 0.195 = 2.552 kg;
6 springs = 1.900 + (0.150 x 3) + 0.157 + 0.195 = 2.702 kg;
7 springs = 1.900 + (0.150 x 4) + 0.157 + 0.195 = 2.852 kg;
8 springs = 1.900 + (0.150 x 5) + 0.157 + 0.195 = 3.002 kg;

b) Determination of L (weight per linear metre of the complete chute)

- 0.9 m2 of stainless steel sheet no. 20 = 7.281 kg;
- 2 m of aluminium angle 2" x 2" x 1/4" = 3,280 kg;
- 2 plates for fixing the rail to the chassis as per fig. 8, weight 1,800 kg (area of the plate times the specific weight of the stainless steel);
- screws, nuts, etc. for fixing the previous plate equal to 0.200 kg (for 10 screws);
$$L = 7.281 + 3.280 + 1.800 + 0.200 = 12.561 \text{ kg}$$

- Calculation of X (distance between cleats);

The table below shows the alternative "X" values for springs with thicknesses of 3/16" and 1/4:

Table 1: M and N values

N	No. of springs	2	3	4	5
M	3/16 spring	7,470	11,205	14,940	18,675
M	spring 1/ 4"	15,390	23,085	30,780	38,475

According to information from Imperial Tobacco Co. of Canada Ltd. the distance between cleats (X) must not be greater than 1.20 metres and not less than 0.60 metres. If the distance is less than 0.75 metres, the drive can only be placed at the end. For 3/16" springs:

N	2	3	4	5
M.N	7,470 kg	11.205kg	14,940 kg	18.675 kg
L	12,561 kg	12,561 kg	12,561 kg	12,561 kg
Q	2,552 kg	2,702 kg	2,852 kg	3.002 kg
X	0,391 m	0,677 m	0,962 m	1,248 m

Conclusion: The accepted solution is 3 or 4 springs per cleat. The solution with 2 and 5 springs per cleat is not accepted.

For 1/4" springs:

N	2	3
M.N	15,390 kg	23,085 kg

L	12,561 kg	12,561 kg
Q	2,552 kg	2,702 kg
X	1,022 m	1,622 m

Conclusion: The accepted solution is 2 springs per cleat. The solution with 3 springs or more per cleat is not acceptable because the distance between cleats is greater than 1.20 metres.

The three possible solutions are listed below:

Alternatives	No. of springs	Distance between cleats
1	2 x 1 /4" springs	1,022 m
2	3 3/16" springs	0,677 m
3	4 x 3/16" springs	0,962 m

We have the following considerations to make:

- Alternative 2 does not allow an intermediate drive (less than 750 mm);
- Alternative 3 is more expensive than Alternative 1ª and Alternative 2ª (more springs and cleats).

We therefore conclude that alternative 1 is the most suitable, as well as being close to 1.00 m, which is the one most commonly used by experience, as shown in the figure below.

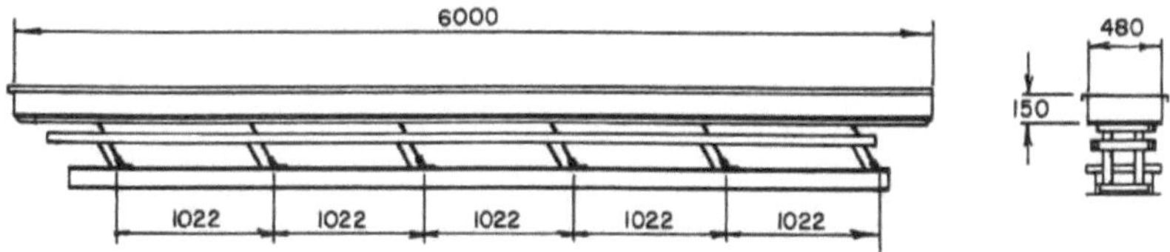

5.9.1.4 - Counterweight calculation (bar thickness and length)

a) Total weight to which all the chassis springs are subjected:
 - 5 complete single aluminium cleats: 5 x 2.552 kg = 12.760 kg;
 - 1 complete cast iron drive cleats: 1 x 3,300 kg = 3,300 kg;
 - 6 m of complete rail (L): 6 x 12.561 kg = 75.336 kg;

Total = 12,760 + 3,300 + 75,336 = 91,396 kg

b) Partial weight to which the counterweight springs are subjected, not including the rectangular bars:

- 5 simple cast iron cleats: 5 x 5.60 = 28,000 kg;
- 1 complete cast iron drive cleats; 6,300 kg;

- screws, nuts and washers for fixing the cleats: 6 x 0.150 = 0.900 kg;

Total: 28,000 + 6,300 + 1,800 + 0,900 = 37,000 kg

So the weight of the (exact) counterweight bar must be the difference between these 2 values:

$$91,396 - 37,000 = 54,396 \text{ kg}$$

The length of the counterweight bar can vary between a maximum and minimum value: the maximum being 0.200 metres in overhang on each side (it cannot be longer due to vibration problems) and the minimum being 0.063 metres, which is the bar placed next to the cleat's ear:

Minimum length = 6 x 1.00 + 0.126 = 12.25 m
Maximum length = 6 x 1.00 + 0.400 = 12.80 m

Therefore, the ideal weight of the counterweight bar can vary between: 54.396/12.25 and 54.396/12.80, i.e. 4.44 kg and 4.23 kg; in the table of weight per linear metre we find the intermediate value of the 7/16" bar with 4.43 kg and 2 bars with 6.14 kg will be used.

Obviously, this value is the exact calculation value, but in practice there are weight differences that must be corrected during the fixing of these bars. Further on in item 6.8, we indicate the steps that must be taken during manufacture to correct these parts.

5.9.1.5 - Drive location

As any of the 3 types of drive can be used due to the distance between cleats, we must analyse which is the best alternative from the point of view of maintenance and the layout to be adopted.

5.9.1.6 - Determining the electric motor

The conveyor will have 12 springs and in this case the table in chapter 5 item 5.7 indicates that a 1 HP motor should be used.

5.9.1.7 - Manufacturing drawing

Figure 37 shows the sketch of the vibrating conveyor for manufacturing.

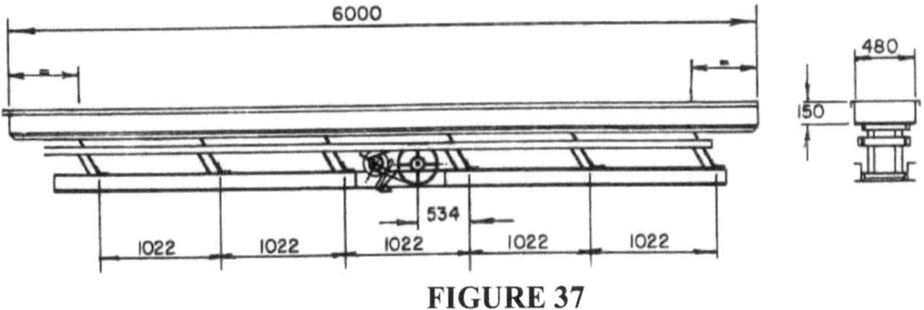

FIGURE 37

5.9.2 - Case 2

Design 1 vibrating conveyor 5.5 metres long, inclined at 10° to the horizontal, to transport threshed tobacco with 18% humidity, requiring a product flow of around 1,800 kg/h.

5.9.2.1 - Choice of rail size (width) and eccentric rpm.

It is recommended that the design flow be 30 per cent higher than the predicted flow. So the design flow will be 2,340 kg/h. From experience, we initially decided to use a 620 mm wide chute with a height of 150 mm (100 metre smoke belt). We set the eccentric speed at 480 rpm (values chosen hypothetically, but based on our experience); the flow rate will then be calculated as follows:

$Q = A \times B \times P \times V$
A = 0.620 m (gutter width);
B = 0.100 m (height of the smoke mat);
$P = 43.2 \text{ kg/m3}$ (specific weight of the material - graph 4);
V = 895.72 m/h (transport speed - graph 7);
$Q = 0.620 \times 0.100 \times 43.2 \times 895.72 = 2,400 \text{ kg/h}$

Therefore, the chosen channel of 620 mm by 150 mm and the eccentric rotation of 480 rpm meets the flow requirements of the project.

5.9.2.2 - Determining the weight each spring is subjected to

With this eccentric value already determined, we go into graph 1 and find 2 values for the weight per spring corresponding to the 3/16" and 1/4" eccentric. As both are less than 8 kg per spring, we must analyse the economic aspect and the values we will find for the distances between cleats (X).

5.9.2.3 - Calculating the distance between cleats (X)

$M. N = L . X + Q$; (See details in item 5.4)

We have the following alternatives to calculate:

M1 = 3.2 kg/spring for springs with a thickness of 3/16";
M2 = 6.1 kg/spring for springs with a thickness of 1/4";
N1 = 2 springs;
N2 = 3 springs;
N4 = 4 springs;
N5 = 5 springs;

a) Determination of Q (weight of a complete aluminium cleat):

- 1 aluminium cleat for 620 mm rail = 2,500 kg;
- 1 spring fastener set on cleat = 0.150 kg;
- Plate for fixing the rail to the cleat, (calculate the area of the plate and multiply by the specific weight of stainless steel) = 0.252 kg;
- The weight of 1 1/4" x 5/8" screw for fixing the aluminium cleat to the chassis is 0.015 kg; as an average of 13 screws are used, the total weight is 0.195 kg;

We then have the following alternatives for the value of M:

2 springs = 2.500 + (0.150 x2   ) + 0.252 + 0.195=3 .247kg   ;
3 springs = 2.500 + (0.150 x3   ) + 0.252 + 0.195=3 .397kg   ;
4 springs = 2.500 + (0.150 x4   ) + 0.252 + 0.195=3 .547kg   ;
5 springs = 2.500 + (0.150 x5   ) + 0.252 + 0.195=3 .697kg   ;

b) Determination of L (weight per linear metre of the complete chute)

- 1.22 m2 of stainless steel sheet No. 20 = 9.869 kg;
- 2 m of aluminium angle 2" x 2" x 1/4" = 3,280 kg;
- 2 plates for fixing the rail to the chassis as per fig. 8, weight 1,800 kg (area of the plate times the specific weight of the stainless steel);
- screws, nuts, etc. for fixing the previous plate equal to 0.200 kg (for 10 screws;

$$L = 9.869 + 3.280 + 1.800 + 0.200 = 15.149 \text{ kg}$$

c) Calculation of X (distance between cleats)
Using the formula M.N. = L. X + Q , we have several alternatives for M and N as shown in the table below:

Table 1: M and N values

N	No. of springs	2	3	4	5
M	3/16" spring	6,4	9,6	12,8	16,0
M	spring 1/ 4"	12,2	18,3	24,4	30,5

According to information from Imperial Tobacco Co. of Canada Ltd. the distance between cleats (X) must not be greater than 1.20 metres and not less than 0.60 metres. If the distance is less than 0.75 metres, the drive can only be placed at the end.

For 3/16" springs:

N	2	3	4	5
M.N	6,400kg	9,600kg	12,800kg	16,000kg
L	15.149kg	15.149kg	15.149kg	15.149kg
Q	3.247kg	3.397kg	3.547 kg	3.697 kg
X	0,210 m	0,410 m	0,610 m	0,810 m

Conclusion: The accepted solution is 5 springs per cleat. The solution with 2, 3 and 4 springs per cleat is not accepted.

For 1/4" springs:

N	2	3
M.N	12,200kg	18,300kg
L	15.149kg	15.149kg
Q	3.247kg	3.397kg
X	0,59 m	0,980 m

Conclusion: The accepted solution is 3 springs per cleat. The solution with 2 springs is not acceptable because the distance between cleats is less than 0.60 metres.

The three possible solutions are listed below:

Alternatives	No. of springs	Distance between cleats
1	4 x 3/16" springs	0,61 m
2	5 3/16" springs	0,81 m
3	3 1/4" springs	0,98 m

We have the following considerations to make:

The 2nd and 3rd alternatives can be adopted, but we favour the 2nd alternative;

- Alternative 1 does not allow an intermediate drive (less than 750 mm), so we shouldn't adopt it, especially since the conveyor is inclined.

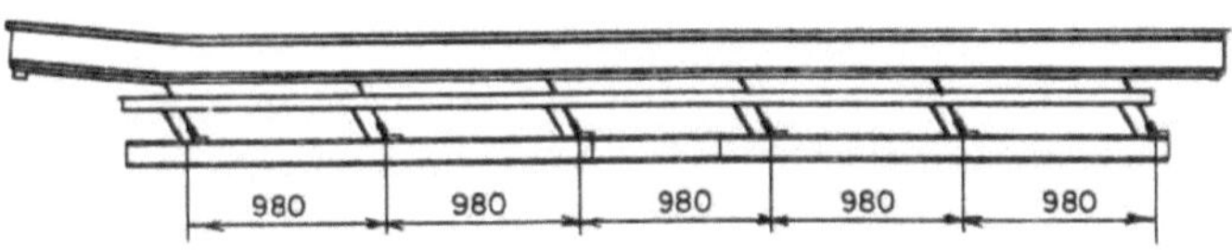

In this case, one end of the overhang will be 0.50 metres and the other 0.60 metres.

5.9.2.4 - Counterweight calculation (bar thickness and length)

-) Total weight to which all the chassis springs are subjected:
- 4 complete single aluminium cleats: 4 x 3.397 kg = 13.588 kg;
- 1 complete cast iron drive cleats = 3,300 kg;
- 5.5 m complete chute (L): 5.5 m x 15.149 kg = 83.32 kg;
Total = 13.588 + 3.300 + 83.32 = 100.21 kg;

-) Partial weight to which the counterweight springs are subjected, not including the rectangular bars:
- 4 simple cast iron cleats: 4 x 5.60 = 22.40 kg;
- 1 complete cast iron drive cleats; 1 x 6,300 kg = 6,300 kg;
- 5 bolts, nuts and washers for fixing the springs to the cleats = 1,500 kg;
- 5 screws, nuts and washers for fixing the cleats to the bars = 0.750 kg; Total:
22.400 + 6.300 + 1.500 + 0.750 = 30.950 kg;

So the exact weight of the counterweight bar must be the difference between these 2 values:
100.21 - 30.95 = 69.26 kg

Maximum weight = 69.26 / 2 (4 x 1.10 + 0.126) = 69.26/9.04 = 7.66 kg/m
Minimum weight = 69.26 / 2 (4 x 1.10 + 0.400 = 69.26/9.6 = 7.21 kg/m

In the weight per linear metre table (item 5.5, item 1), we find between these values of 7.66 and 7.21, with an intermediate value of 7.59 kg/m, which corresponds to the %" bar.

So the %" bar to be used must measure 69.26/7.59 = 9.12 metres. Each bar must then measure 4.56 metres.

Obviously, this value is the exact calculation value, but in practice there are weight differences that must be corrected during the fixing of these bars. Further on in item 6.8, we indicate the steps that must be taken during manufacture to correct these parts.

5.9.2.5 - Drive location

The type I drive must be used, as shown in fig. 31.

5.9.2.6 - Determining the electric motor

As the conveyor is inclined, the number of springs + 10 per cent must be taken into account.
As such, this figure is in the range for using a 2 HP motor.

5.9.2.7 - Manufacturing drawing

Figure 38 shows the sketch of the vibrating conveyor for manufacturing.

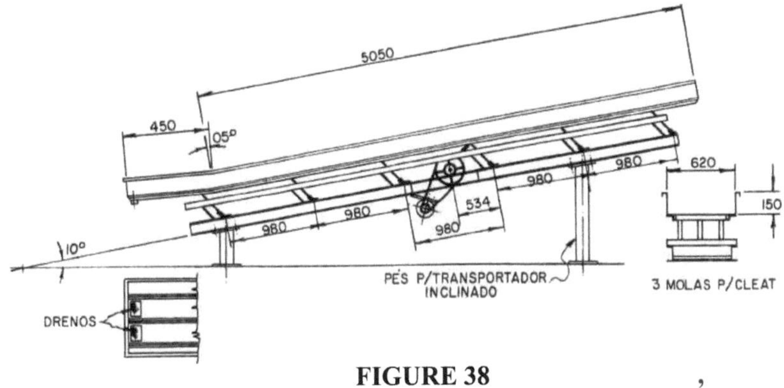

FIGURE 38

Below is the manufacturing sequence for vibratory conveyors, with the care required and the necessary adjustments.

CHAPTER 6

MANUFACTURING

6.1 - Care must be taken when welding in order to avoid as much as possible deformation in the surroundings, which are points of noise generation;

6.2 - Prepare the 2" x 2" x1/4" aluminium angles;

6.3 - Mount the aluminium cleats on the angle brackets;

6.4 - Mount the angles with the cleats on the rail;

6.5 - Weigh the rail with the aluminium angles and cleats;

6.6 - Prepare the counterweight bars;

6.7 - Mount the cast iron cleats on the counterweight bars;

6.8 - Weigh the counterweight bars with the cleats. This weight should be equal to that determined in item 6.5; if it is less, weld small pieces of bar along the rectangular bars to equalise the weights; if it is greater, cut off equal pieces at the ends of these bars;

6.9 - The iron profiles must not be warped;

6.10 - Mount the eccentric shaft with its bearings and connecting rods on the base; it is essential to mark the rpm on the end of the eccentric shaft;

6.11 - Mount the steel springs with the same dimensions as the fibreglass springs on the counterweight; mount the end of the steel springs on the support cleats of the counterweight;

6.12 - Mount the counterweight with the cleats on the base, without fixing it;

6.13 - Mount steel springs with the same dimensions as the fibreglass springs on the rail; mount the rail support cleats at the end of the steel springs;

6.14 - Position the rail on the table; the distance from the centre of the eccentric shaft to the iron cleats supporting the rail, which are connected to the drive cleats, must be 534 mm, as shown in figure 39;

6.15 - Weld the iron cleats supporting the rail to the base;

6.16 - Position the counterweight on the base; the distance between the iron cleats supporting the rail and the iron cleats supporting the counterweight must be 22 mm, as shown in figure 40;

6.17 - Drill the holes for fixing the counterweight support cleats;

6.18 - Assemble the chute and counterweight drive springs with the spring bracket;

6.19 - Fit the regulator screw to the connecting rod and the rail drive spring bracket;

6.20 - The centres of each connecting rod are marked on the end of the eccentric shaft; turn the adjusting screw, causing the eccentric shaft to rotate, in order to position it at the neutral point; The shaft is at the neutral point when the line joining the two centre points of the connecting rods is 30° from the vertical, coinciding with the inclination of the rail support springs, which are also 30° from the vertical; Once these conditions have been verified, secure the adjusting screw with the lock nuts;

For an illustration, see figure 41;

6.21 - Fit the adjusting screw to the connecting rod and the counterweight drive spring bracket and secure it with the locknuts;

6.22 - Replace the steel springs with fibreglass springs in an alternating sequence;

6.23 - Mount the electric motor bracket, electric motor and eccentric shaft pulley;

6.24 - Mount the variable speed drive assembly; gradually vary the speed until you reach a range where you feel the conveyor is working perfectly; this speed is close to the speed calculated for balancing the conveyor;

6.25 - Make the electric motor pulley for the speed found with the variator;

6.26 - Mount the electric motor pulley, "U" chain and guard;

6.27 - Put the following on the conveyor manufacturing plate:

- Conveyor drawing no;
- Weight of chute and conveyor;
- rpm of the eccentric;
- date of manufacture;

6.28 - If the conveyor is to be operated at an incline with

6.29 compensation, switch off the drive by removing the drive springs;

6.29 - With the conveyor horizontal, mark a vertical line on the side of the chute, counterweight bar and base;

6.30 - Tilt the conveyor to the position it will work in;

6.31 - Check the vertical line marking on the rail, counterweight and base, which will no longer match;

6.32 - Adjust the compensation springs until the vertical markings coincide;

6.33 - Connect drive springs;

6.34 - Return to item 6.23 and proceed to item 6.27;

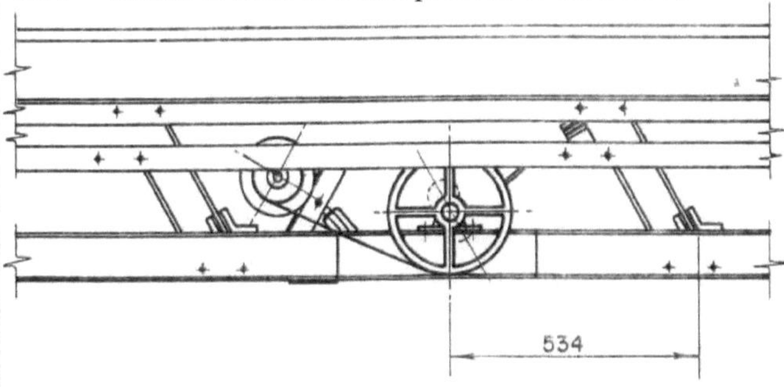

FIGURE 39

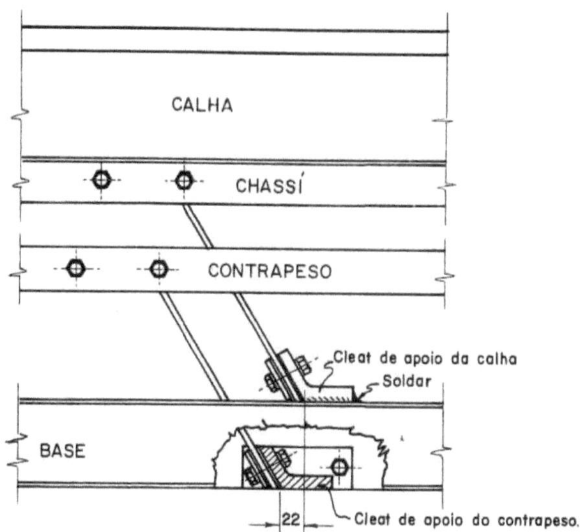

FIGURE 40

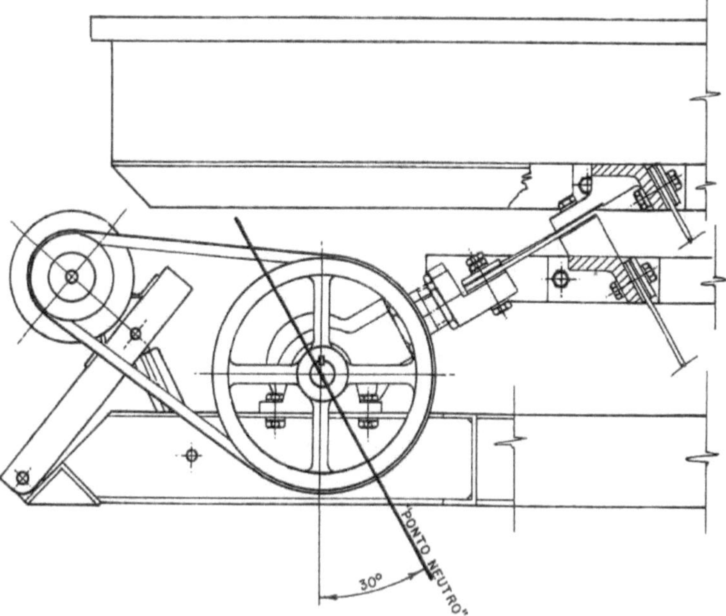

FIGURE 41

CHAPTER 7

INSTALLATION

Care must be taken when transporting vibratory conveyors from the manufacturer to the installation site to avoid mechanical damage. Longer conveyors require greater care when being transported and installed. Sufficient support points must be provided so as not to cause warping of the iron "U" base.

In the case of vibrating conveyors, it is more usual to use steel pipe posts with a diameter of 6" as a support. Figures 42 and 43 show the post designs for horizontal and inclined conveyors.

A post or support should always be installed under the eccentric shaft of the conveyor.

On long conveyors with tall poles, care must be taken with the horizontal force components.

In these cases, 45° arms made of 2" steel tube are usually installed on the end posts, as shown in figure 44.

On vibratory conveyors with 480 mm and 620 mm wide chutes, posts are installed with a spacing equal to or less than 3.5 metres. On heavier conveyors, the spacing must be less than 3.5 metres.

Before permanently attaching a vibratory conveyor to the posts, check that the "U" base is levelled in accordance with the design. If not, correct the levelling with shims. Electrical installations must comply with ABNT technical standards.

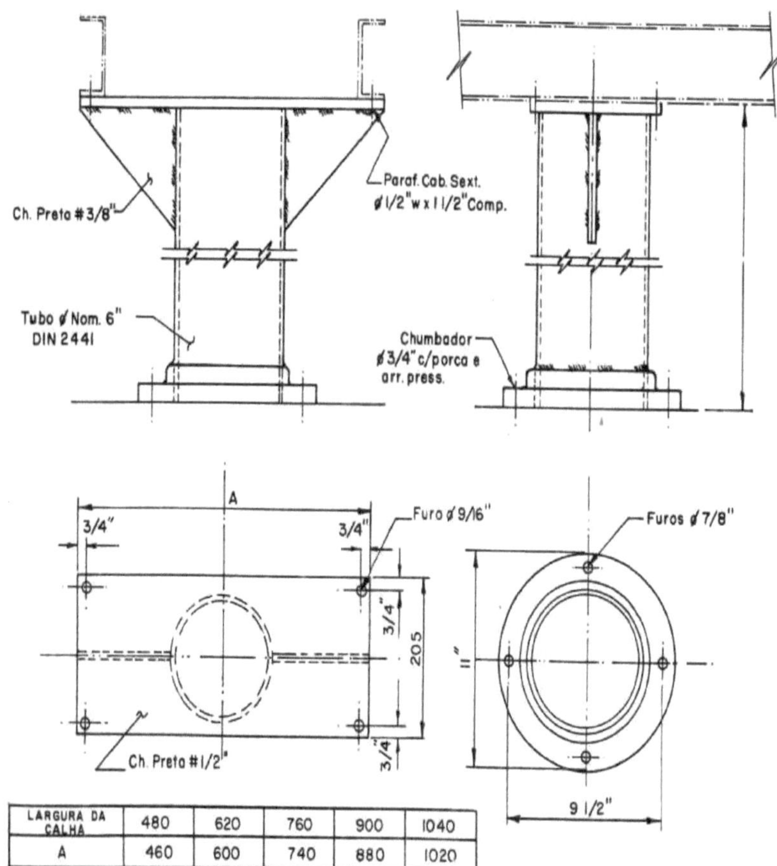

LARGURA DA CALHA	480	620	760	900	1040
A	460	600	740	880	1020

FIGURE 42

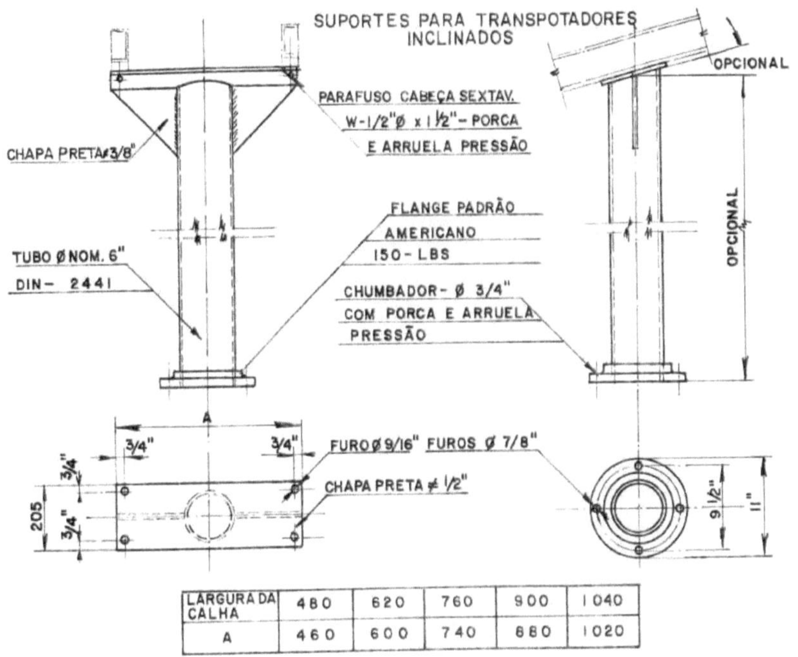

LARGURA DA CALHA	480	620	760	900	1040
A	460	600	740	680	1020

FIGURE 43

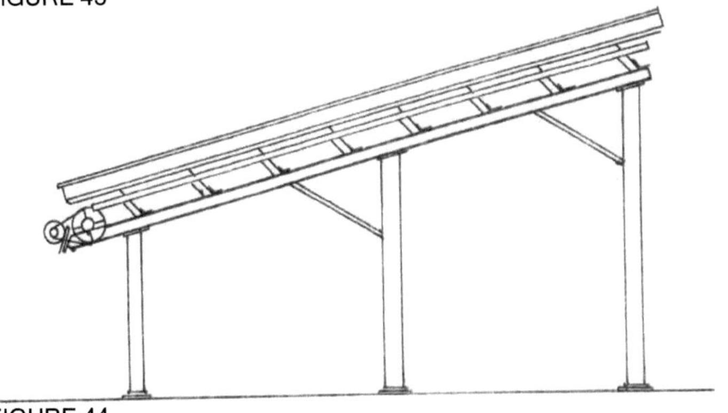

FIGURE 44

CHAPTER 8

MAINTENANCE

Subsequently, the following inspection should be carried out every 3 months:
- General tightening of bolts and nuts;
- Check the fibreglass springs;
- Check the chassis and counterweight drive cleats and the drive springs;
- Tighten the hexagonal nuts on the drive regulator screw;
- Examine the condition of the rolling bearings through noise and temperature;
- Apply grease to the bearings;
- Check the V-belt, tensioner, tensioner lock and pulleys.

MISCELLANEOUS

Imperial Tabacco has built conveyors for 10,000kg/h of tobacco without taking into account the effects of the weight of the product on the balance. This is why the weight of the product is not taken into account in the design.

When carrying out a project, the conveying speed must be determined experimentally on an existing conveyor.

REFERENCES

Dube, J.T.; Manual "Balanced Vibrating Conveyors"; The Imperial Tobacco Co. of Canada Ltd. Montreal, Engineering Departement; Que. p.66.
Chagas, J.S.; Odorico, S.; Manual Transportes Vibratórios; Cia. Souza Cruz, Departamento de Engenharia; Rio de Janeiro; March 1978; p.80.

Printed by Books on Demand GmbH, Norderstedt / Germany